Eco-Restoration of Mine Land

Eco-Restoration of Mine Land

Vimal Chandra Pandey
Babasaheb Bhimrao Ambedkar University, Lucknow, India

Jitendra Ahirwal
Department of Forestry, Mizoram University, Aizawl, India

Roopali Roychowdhury
Department of Biotechnology, Techno India University, Kolkata, India

Ritu Chaturvedi
Department of Botany, St. John's College, Agra, India

WILEY

This edition first published 2023

Registered Office
John Wiley & Sons, Inc., 111 River Street, Hoboken, NJ 07030, USA

Editorial Office
Boschstr. 12, 69469 Weinheim, Germany

For details of our global editorial offices, customer services, and more information about Wiley products visit us at www.wiley.com.

Wiley also publishes its books in a variety of electronic formats and by print-on-demand. Some content that appears in standard print versions of this book may not be available in other formats.

Library of Congress Cataloging-in-Publication Data applied for
Hardback ISBN 9781119872252

Cover image: Wiley
Cover design by Courtesy of Vimal Chandra Pandey and Jitendra Ahirwal

Set in 9.5/12.5pt STIXTwoText by Straive, Pondicherry, India

Contents

Foreword

Environmental degradation by mining is a major concern worldwide. Mining activities have led to the formation of dumps and abandoned mine lands which have negative impacts on the environment immediately and even after several years. Mine lands pose environmental risk due to the presence of moving and transforming heavy metals and metalloids. Several risk assessment methods via calculation of pollution indices have been designed to perform an extensive geochemical assessment to analyze soil quality of mine lands, to carry out ecological and human health risk assessment, to determine the origin of natural or anthropogenic pollutants, and to study the degree of soil degradation. Reversing soil degradation, regaining soil fertility, and remediation of polluted soil are the major challenges and need redressal immediately after mining operations. Revegetation emerges out to be a feasible solution for ecosystem restoration. To overcome this challenge, the book addresses the scope of mine land restoration for achieving the United Nations Sustainable Development Goals.

This book is useful to select specific plant species according to site specificity of the mined land. As this book explores, there are clearly some potential opportunities in mine land restoration. This book provides a concept of ecological restoration of mined land for regaining ecosystem services and raising livelihoods. It is also the intention of this book to enable the reader to become better informed with the mechanisms of plant's adaptation to adverse conditions prevailing on mine land. General public has become more aware of the importance of preserving the environment and human health through eco-restoration of mine land. This authored book is well timed and has up to date information that offer a cutting-edge synthesis of scientific, recent experiential, and established knowledge as a single source on different aspects of mine land restoration.

I congratulate the authors, Dr Vimal Chandra Pandey, Dr Jitendra Ahirwal, Dr Roopali Roychowdhury, and Dr Ritu Chaturvedi, for bringing out this valuable book published by a renowned publisher Wiley. The book consists of six chapters covering various aspects of *Eco-Restoration of Mine Land*. I believe this book will be a notable asset for mining industry professionals, eco-engineers, eco-planners, researchers, PhD scholars, plant scientists, environmentalists, practitioners, entrepreneurs, policy makers, and other stakeholders alike.

28 June 2022

Dr Jan Frouz
Professor and Director
Environmental Centre, Charles University, Prague,
Czech Republic

Preface

Mining is an integral part of the national economy. The excavation of land resources is essential to fulfilling energy demands but causes multitudinous effects on the environment. The global mining market is expected to grow at a rate of 12% per annum which can further imperil ecological processes and functions. Asia Pacific is the largest mining market accounting for 70% of the global mining activities followed by North America and Western Europe. The mining of minerals, coal, and metals such as iron, copper, zinc, lead, and nickel are dominant in mining sectors. However, coal, anthracite, and lignite mining alone contributed to more than half of the mining markets. Correspondingly, land degradation due to mining activities is increasing at an alarming rate. Thus, repairing these degraded lands is essential to reinstate ecosystems and becomes a part of sustainable mining strategies.

Surface mining leads to changes in geological settings and removal of vegetation cover, alters soil properties, and degrades large pieces of land. Mining is site specific, and land degradation due to mining is highly dependent on the stripping ratio – ratio of ore extracted to overburden materials. These activities are also responsible for relocating indigenous populations that affect their livelihood and alter land use patterns in many parts of the world. Mining is therefore responsible for environmental, social, and economic transformation. Land degradation and environmental pollution caused by mining activities have attracted a great deal of scientific and political attention. Several restoration efforts have been taken to restore the mine-degraded lands, and most of them are based on reforestation and afforestation. These revegetation activities improve soils, biogeochemical cycles, and biodiversity; reduce toxicity; and maintain the overall aesthetic values of the post-mining lands.

Eco-Restoration of Mine Land is focused on a continuum of restoration activities namely environmental, social, and economic in terms of ecological restoration that is primarily based on post-mining lands. This textbook is a compact and precise collection of relevant literature and scientific practices designed to combat land degradation and biodiversity losses, restore soil fertility, and phytomanagement of mine waste. In general, ecological restoration is the process of assisting the recovery of any degraded, damaged, and destroyed ecosystem to bring it back to its original conditions. This attempts to reinstate species composition, ecosystem productivity, and structure prior to disturbance. This textbook comprises different chapters that describe basic theory of restoration science and numerous facts and findings related to the ecological restoration of mined lands. Mining and its overall impact on the environment, soil contamination, and its phytomanagement, different bio-geotechnologies employed to improve ecosystem structure and functions such as soil fertility and carbons sequestration, and how to assess the impact of restoration practices by selective indicator parameters are the major focus of this book. Besides, how a restoration of mine degraded lands provides ecosystem services such as supporting, regulatory, provisioning, and cultural are described in brief. How mine restoration is helpful in achieving the United Nations sustainable development goals and its link to the circular economy are also discussed.

Eco-Restoration of Mine Land addresses the basis theory and practices of ecological restoration of mine degraded land. This book is a collection of specific literature that comprehensively deals with the impact of mining on the environment and human health, techniques of mine restoration, ecosystem services including carbon sequestration provided through mine restoration, metal toxicity reduction by plants, soil fertility management, and sustainable mining practices. This book is designed for a restoration practitioner working in different mining areas. This book will be helpful to students belonging to different fields of applied ecology and environmental sciences such as restoration ecology, earth sciences, ecotoxicology, environmental engineering, microbiology, and environmental pollution. This book will also serve as a reference guide for policy planners, land managers, and restoration ecologists to manage post-mining lands. Finally, we believe this book will integrate environmental, biological, engineering, ecological, social, and economic aspects of ecosystem management and thereby assist readers to develop multidisciplinary approaches to restoring mine degraded lands.

Vimal Chandra Pandey
Babasaheb Bhimrao Ambedkar University, Lucknow, India
Jitendra Ahirwal
Mizoram University, Aizawl, India
Roopali Roychowdhury
Techno India University, Kolkata, India
Ritu Chaturvedi
St. John's College, Agra, India

Acknowledgments

We sincerely thank *Rituparna Bose* (editor), *Mandy Collison* (managing editor), and *Frank Weinreich* (publisher) from Wiley for their excellent support, guidance, and coordination during the production of this fascinating project. We thank all the reviewers for their careful and insightful review of the book chapters. The authors are highly thankful to Professor Jan Frouz, Director, Environmental Centre, Charles University, Prague, Czech Republic for writing the foreword at short notice. We would like to thank our respective families for their endless support and encouragement.

About the Authors

Dr Vimal Chandra Pandey featured in the world's top 2% scientists curated by Stanford University, United States. Dr Pandey is a leading researcher in the field of environmental engineering, particularly phytomanagement of polluted sites. His research focuses mainly on the remediation and management of degraded lands, including heavy metal-polluted lands and post-industrial lands polluted with fly ash, red mud, mine spoil, and others, to regain ecosystem services and support a bio-based economy with phytoproducts through affordable green technology such as phytoremediation. Dr Pandey's research interests also lie in exploring industrial crop-based phytoremediation to attain bioeconomy security and restoration, adaptive phytoremediation practices, phytoremediation-based biofortification, carbon sequestration in waste dumpsites, climate resilient phytoremediation, fostering bioremediation for utilizing polluted lands and attaining UN Sustainable Development Goals. His phytoremediation work has led to the extension of phytoremediation beyond its traditional application. He is now engaged to explore profitable phytoremediation with least risk, low input, and minimum care. Dr Pandey worked as a CSIR-Pool Scientist and DS Kothari Postdoctoral Fellow at Babasaheb Bhimrao Ambedkar University, Lucknow; a consultant at the Council of Science and Technology, Uttar Pradesh; and DST-Young Scientist at CSIR-National Botanical Research Institute, Lucknow. He is the recipient of a number of awards/honors/fellowships and is a member of the National Academy of Sciences India. Dr Pandey serves as a subject expert and panel member for the evaluation of research and professional activities in India and abroad for fostering nature sustainability. He has published over 100 scientific articles/book chapters in peer-reviewed journals/books. Dr Pandey is also the author and editor of 10 books published by Elsevier and CRC Press, with several more forthcoming. He is an associate editor of *Land Degradation and Development* (Wiley); an editor of *Restoration Ecology* (Wiley); an associate editor of *Environment, Development and Sustainability* (Springer); an associate editor of *Ecological Processes* (Springer Nature); an academic editor of *PLOS ONE* (PLOS); an advisory board member of

Ambio (Springer); an editorial board member of *Environmental Management* (Springer), *Discover Sustainability* (Springer Nature), and *Bulletin of Environmental Contamination and Toxicology* (Springer). He also works/ worked as guest editor for many reputed journals.

Dr Jitendra Ahirwal is a postdoctoral researcher in the Department of Forestry, Mizoram University, Aizawl, India. Dr Ahirwal received his PhD on the topic "Reclaimed mine soil quality and carbon sequestration potential of mining induced land use changes in a dry tropical climate" from Indian Institute of Technology (Indian School of Mines), Dhanbad, India. His main research areas are ecological restoration, carbon sequestration, land use change, CO_2 flux, and climate change. He has published over 30 research papers in reputed journals and several book chapters with leading international Publishers. Dr Ahirwal received the Presidential International Fellowship Initiative (PIFI) award for postdoctoral research from the Chinese Academy of Sciences at CAS-Institute of Botany, Beijing, China. Dr Ahirwal serves as an active reviewer for several reputed international journals and a member of Society for Ecological Restoration (SER), USA.

Dr Roopali Roychowdhury is currently engaged in bioremediation and phytoremediation of polluted sites as well as in deciphering the mechanisms of enhanced bioremediation. She obtained her PhD in environmental biotechnology from the Department of Biotechnology, Techno India University, West Bengal, India. She has published 22 research papers in reputed international and national journals and 3 book chapters.

Dr Ritu Chaturvedi recently received her PhD degree on the topic "Risk Assessment and Phytoremediation Efficiency of AMF-assisted Vegetable Plants under Heavy Metal Stress" working in the Department of Botany, St. John's College, Agra, Uttar Pradesh, India. Her early research was focused on bioremediation of heavy metal through fungi and EDTA. She is also working on plant-based bioremediation or phytoremediation. Till now she has published 6 research papers in reputed international journals, 1 review article and 6 book chapters with Springer and Elsevier Publishers. She also serves as a potential reviewer for several reputed national and international journals.

1

Mine Land and its Environmental Impacts

Abstract

Exploration of mineral resources and mining is indispensable for the economic development of any nation, but its ruinous impact on environment is inevitable. Though mining sector significantly contributes to gross domestic product (GDP) and employment of workforce in any nation, it causes irreversible harm to the environment. This sector is known to afflict environment by causing disturbance in air quality, pollution of water bodies, depletion of soil fertility, desertification, deforestation, endangerment of wildlife, etc. Therefore, any mining project must necessarily include a mine reclamation closure plan in order to exploit resources judiciously, not at the cost of environment. This chapter highlights the environmental impacts of mining and presents the solutions that can be adopted with mining to protect the environment.

Keywords *Mining; pollution; ecological disruption; acid mine drainage; reclamation*

CONTENTS

Eco-Restoration of Mine Land, First Edition. Vimal Chandra Pandey, Jitendra Ahirwal, Roopali Roychowdhury, and Ritu Chaturvedi.

1.1 Introduction

Mineral wealth accounts to more than 70% of resources required worldwide (Pashkevich 2017). Mining activities pertain to extraction of nonrenewable resources (fuels, metallic, nonmetallic, and minor minerals) from a resource-rich ore. Mining resources are pivotal to promote economic growth and expansion of any nation. In India, mining industry contributes around 10–11% to gross domestic product (GDP) of total industrial sector. Around 80% of mining in India pertains to coal, and the rest is related to remaining minerals. India is the largest producer of mica blocks and mica splitting and produces ~60% of world's mica. As of 2019, India was fourth largest producer of iron ore and chromium; fifth largest producer of bauxite, zinc, and graphite; seventh largest producer of manganese, lead, and sulfur; and eleventh largest producer of titanium and uranium worldwide (Corporate Catalyst India [CCI] 2010). Post announcement of New Mineral Policy in 1993, mining sector in India was opened to foreign direct investment (FDI), but the investment remained low due to restrictions in obtaining approvals (Mehta 2002). Since then the FDI policy in the mining sector has been gradually liberalized, for example, by introducing automatic approval route, and has remained successful to attract FDI. However, there are certain challenges that need to be addressed in mining sector such as upgrading and adapting to eco-friendly technologies, curtailing environmental degradation and post-mine rehabilitation, and addressing social issues dealing with displacement and marginalization of native dwellers and its economic consequences. In India, premier organizations such as Geological Survey of India (GSI) and Mineral Exploration Corporation Ltd (MECL) carry out survey and mineral exploration activities, while Indian Bureau of Mines (IBM) is engaged in regulation and conservation. Though IBM updates the inventory of mineral deposits every five years, immense research and exploration still need to be carried as India has diverse mineral deposits yet to be discovered (Kumar 2019).

Environmental damage and mining scale do have an unequivocal relationship (Li et al. 2016). Around 90–95% of extracted mineral resources are lost as wastes, which result in the formation of technogenic dumps (Pashkevich and Petrova 2015). It is well documented that both active and abandoned mine sites affect terrestrial (soil fertility and erosion), aquatic (water quality and biota), and atmospheric (air quality) components of ecosystem adversely (Favas et al. 2018). It becomes alarming when active mine sites or abandoned mine lands are categorized as devastated landscape, where the natural tendency of auto regeneration is lost and only human intervention can lead to reclamation. Large-scale mining projects involve more technology, generate more wastes, and run for longer time, i.e. exposing environment to more damaging effects. Moreover, the topography and climate of a mine site also govern the level of environmental damage that is bound to occur. The mobilization of heavy metals (such as Cd, Cu, Hg, Mn, Ni, Pb, and Zn) and metalloids (As and Sb) during mining through extraction from ores and processing releases these toxic elements into the environment. Precipitation, temperature, wind flow, humidity, and other factors strongly determine the extent of diffusion of pollutants to surrounding environment. The atmosphere plays key roles in dispersion of air pollutants, whereas precipitation influences spread of water pollutants (Li et al. 2016).

Different mining projects vary in their plan according to the physical and chemical properties of the ore being mined. A proposed mining project begins with an explorative phase whereby the extent and value of the ore are evaluated via surveys, field studies, test drillings, and excavations and concludes with a post-closure phase. Each of these phases can be held responsible for a range of outcomes on environmental health. Even if the findings of explorative phase render the mining project unsuitable, the area is already de-vegetated and profoundly disturbed in the name of exploration. Whereas if the exploration proves that the ore is of sufficient grade and the mining project is economical, further developments (not without adverse impacts) are made including clearing of vast areas and construction of access roads. Thereafter, the active mining phase, i.e. extraction and beneficiation of target metal from its ore, begins. Metallic ores are generally buried under huge amount of waste rock or overburden, which needs to be removed by any of the processes depending on the type of ore. Strip ratio, i.e. the ratio between the amount of overburden generated and the mineral ore, is generally more than one or much higher. The overburden may contain toxic elements which either get deposited on-site or dumped into a landfill. If the ore is buried very deep in the ground, it is accessed through open-pit mining which results in high volume of overburden removal. Environmentally, it is the most destructive mining process especially in tropical forests, as it involves clearing of vegetated areas by cutting or burning native vegetation. Moreover, the constant increase in demand for minerals can be met from open cast mines ultimately leading to massive deforestation. But it is lower in cost and suitable for lower-grade ore mining (Environmental Law Alliance Worldwide 2010).

In another technique, placer mining is used if the metal is associated with stream sediments or floodplains, e.g. gold. Since it takes place within a streambed, it discharges high load of sediments which adversely affect surface water to great extent. An environmentally less destructive way to access ore is through underground mining as it minimally affects landscape, but it is costlier and poses safety risks as compared to open-pit mining. It can affect hydrogeological conditions and may induce earthquakes (Li et al. 2016).

After successful removal of overburden, the ore is extracted using heavy machinery and transported to processing units for beneficiation. Beneficiation includes physical and/or chemical processes to separate metals from nonmetallic components of ore. Even high-grade ores often contain toxic metal(loid)s such as lead, cadmium, and arsenic. Techniques such as milling, magnetic separation, solvent extraction, and leaching are employed for this purpose. However, these processes generate waste in the form of dumps, tailings, and leach materials. Exploitation of different types of minerals results in divergent environmental impacts. Combustible organic materials generate solid wastes and emit toxic gases. Mining of metallic ores leads to heavy metal(loid) pollution (Li et al. 2016). Beneficiation of gold and silver ores requires spraying of cyanide solution over finely milled ore to dissolve target metals and collected thereafter. Likewise, leaching of copper requires sulfuric acid and releases high-volume waste in the form of tailings which need to be addressed for proper disposal in order to prevent mobilization and release of toxic compounds into the environment. The involvement of cyanide calls for special attention for its fate and impact on environment, since it disrupts oxygen metabolism and causes acute toxicity at all trophic levels even at low concentrations and for over long periods (González-Valoys et al. 2022).

Post-mining, the ultimate aim of any mining project should be to reclaim the mining site almost to its pre-mining condition. And hence, mining closure plan must be designed scientifically justifying actions and measures adopted to prevent pollution and the insurance of funds for reclamation.

1.2 Environmental Impacts of Mine Land and Ecological Disruption

Minerals obtained from mine lands are limited and nonrenewable in nature. Due to their applications in various industries, reckless exploitation of these natural resources often leads to environmental imbalances. Mining activities are known

Table 1.1 Effects of mining activities on environment.

Air	Water
• Dust emission due to blasting operations in surface mines • Release of GHGs such as methane and carbon dioxide from coal mines • Release of heavy metal(loid)s and SO_x during smelting operations	• Mining operations require large amounts of water which is sourced from underground or nearby water bodies • SO_x compounds released into air react with water and form sulfuric acid which pollutes groundwater or water bodies
Land	**Life**
• Blasting activities and creation of sink holes affect land integrity • Mining leads to deforestation and loss of biodiversity • Mining leads to soil erosion and contamination, affecting soil quality and fertility adversely	• Underground mining activities can be hazardous due to poor ventilation and visibility • Rock falls and explosions in mining sites are hazardous for all forms of life • Exposure to dust and radiations poses health risks to workers

to impact forest ecology, wildlife, local climate, and water balance of the area. The Sustainable Development Networking Programme, India states how mining activities in a region affect natural resources in an area (Table 1.1). Figure 1.1 illustrates the pathway of receptivity of pollutants from mine sites to plants, animals, and humans.

Mining activities amend the landscape completely to a hostile state and are responsible for multiplex environmental issues. The magnitude of ecological impacts is not just limited to aesthetics and visual features but extend from soil erosion to pollution of air and water, habitat destruction to loss of species, resulting in overall perturbance to environment. These disruptions reduce the range of biological activities in proximity of mine site leading to decline in function and stability of the ecosystem (Li et al. 2016). Mining activities not only lead to pollution but also modify the landscape significantly, and hence disrupting the microclimate of that region.

1.2.1 Soil Erosion

Clearing vegetation and excavations are the first phases of any mining project. Open-pit sites, overburden heaps, tailing piles, dump leaches, etc. contribute chemical pollutants and are prime sources of sediment loading. The eroded sediment as a part of surface runoff often gets deposited in nearby floodplains or loads into surrounding water bodies. The minerals present in contaminated sediments

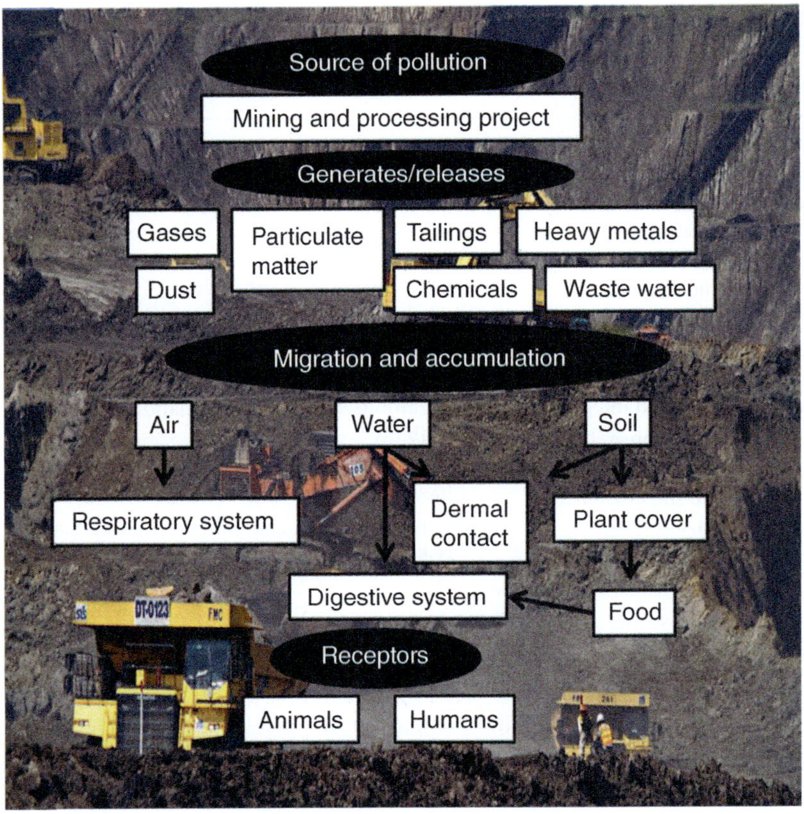

Figure 1.1 Impact of mining on environment.

may lead to reduction in pH of soil as well as water, affecting the physicochemical properties thereby. Lower soil pH renders the soil unsuitable for vegetation and also facilitates mobilization of heavy metals, enabling them leaching into subsoil or groundwater. Increase in particulate matter in water bodies due to surface run-off may prove toxic for the fishes and affect the complete food chain (Geelani et al. 2013; Pashkevich 2017). Erosion followed by sedimentation is responsible for the accumulation of sediments and rock dust in water bodies and floodplains, causing reduction of storage capacity in surface water and modification of aquatic habitat (Barve and Muduli 2011).

1.2.2 Loss of Soil Fertility and Desertification

Mining activities habitually reform the even and tranquil landscape by continuously uncovering mining material. Since a large area is involved under mining, huge quantities of earthen material are exposed on sites especially in case of hard rock mining (Geelani et al. 2013). Mining activities transform the nature of these mine lands and make them less fertile by removing the topsoil layer, leaving them more prone to soil hazards (Buta et al. 2019). Stripping off topsoil and exposing the subsoil layer to heavy equipment and machinery subject them to damage. Operations requiring high temperature render the soil dehydrated, making it dense and compact. Fire used to clear off vegetation also affects minerals present in soil, altering structure and composition. Burning primarily reduces soil's productivity by destroying humus and nutrients like nitrogen and phosphorous. Such soil with altered texture is easy to break and less fertile (Li et al. 2016). Mining wastes and effluents from dumps are highly acidic and their release results in altered pH of soil, ultimately leading to loss of habitat for various microbes and plants (Barve and Muduli 2011). It becomes apparent in the form of reduced crop yield and survival.

1.2.3 Loss of Biodiversity

The very first phase of mining, i.e. exploration, involves clearing of vegetation and removal of topsoil besides releasing pollutants into the soil in upcoming phases. Disturbances caused in the natural habitat due to air, water, and noise pollution lead to displacement of fauna. As we know that habitat loss is the primary cause of species extinction, the extent to which mining activities affect biodiversity can be understood. Habitat fragmentation as a consequence of mining has been shown to affect amphibian diversity and distribution in Western Ghats, India (Krishnamurthy 2003). As plant and animal communities are interdependent and depend on common natural resources which get polluted by mining, entire communities get affected. The massive adverse effect of mining on abiotic factors which is responsible for the change in natural biota can be depicted by low Simpson's diversity index (SDI) in mining areas. Mining activities perturb homeostasis of environmental factors which leads to disturbed ecosystem (Basavaraja and Kumar 2020). Mining activities affect abiotic components of environment directly, which in turn perturb the distribution of biotic components, i.e. natives might be replaced by exotics as a consequence of unplanned mining and rehabilitation. Invasion is of, by and large, occurrence in areas disturbed due to land transformation and anthropogenic activities. Several invasive exotic plant species have been identified in areas undergoing reclamation, which are able to disrupt the growth of other plants either by covering the canopy and competing for sunlight or by possessing deep root systems and competing for nutrient absorption (Yusuf and Arisoesilaningsih 2017). Invasive exotic species often invade recent reclaimed mine sites, if they lie in close proximity to older reclaimed mine lands. Such plants not only compromise revegetation success and ecosystem succession but also are a threat to the existence of native and desirable species (Zipper et al. 2019).

Additionally, increase in concentration of SO_2 in air due to operations such as blasting and combustion of fossil fuels is deleterious to sensitive species. Alteration in species composition leads to reduction in primary productivity and perturbs trophic relationships, hence possesses extensive effects for both floral and faunal communities.

1.2.4 Air Quality

Mining operations namely, excavation, blasting, and erosion mobilize and disperse large amounts of particulate matter into air, which is the primary pollutant of concern. Operation of heavy machinery and vehicles generates large amounts of dust which flows due to high wind velocity. Combustion of fuels by transportation facilities and equipments such as generators and smelting furnace contributes to gaseous emissions (containing volatile organic compounds [VOCs], SO_x, NO_x, CO, and volatile heavy metals) and leads to deterioration of air quality. Air polluted with particulate matter has reduced oxygen saturation and is capable of causing discomfort in living organisms (DeMeo et al. 2004). Polluted air also carries more SO_2 which can inhibit photosynthesis by disrupting photosynthesis mechanism. SO_2 promotes stomatal opening in plants and leads to excessive water loss due to transpiration. It is detrimental to both quality and quantity of plant yield. Another indirect consequence of SO_2 pollution is occurrence of acid rain that is capable of leaching out nutrients from plant canopy and soil (Varshney et al. 1979).

Mercury, generally present in gold ore, gets vaporized during processing operations and gets released in atmosphere. Inhalation of iron and manganese mine dusts for extended durations can lead to siderosis and manganosis, which are occupational diseases causing damage to respiratory system. Pollutants upon entering the environment undergo physical and chemical changes and ultimately impact life forms adversely (Basavaraja and Kumar 2020).

1.2.5 Emission of Greenhouse Gas and Climate Change

Heavy machinery, power generation, and transportation facilities employed during mining operation are known to release massive amounts of gases such as CH_4, CO_2, NO_x, which are detrimental to ozone layer and known as greenhouse gases (GHGs). Some of these gases such as CO_2 and NO_x are responsible for acid rain and also act as secondary pollutants (Li et al. 2016). Increase in the emission of GHGs due to anthropogenic activities is primarily responsible for increased greenhouse effect, i.e. global warming. Melting of glaciers, rising sea level, desertification, etc., are the alarming consequences of global warming which have threatened the existence of life on Earth. Depletion of ozone layer will lead to more UV rays entering atmosphere and reaching Earth's surface, causing magnitude of impact on plants, animals, and humans (Basavaraja and Kumar 2020).

Moreover, mining projects need to be evaluated for their carbon impact as they potentially modify global carbon budget. Life cycle assessment methodology can be used to estimate environmental impacts and has been applied to different mineral processes (Franks et al. 2010). Tropical forests act as lungs for the planet by absorbing CO_2 and releasing O_2, but this equilibrium gets hampered when a proposed mining project demands destruction of forests.

1.2.6 Water Quality Deterioration

Water is the most crucially affected resource during mining operations and is called "mining's most common victim." Mining by its tendency consumes, diverts, and pollutes water resources, and its impacts can be broadly classified in four major types: acid mine drainage (AMD) or acid rock drainage (ARD), processing chemicals discharge, heavy metal pollution and leaching, and erosion and sedimentation (Emmanuel et al. 2018). Large quantities of wastewater containing heavy metals, reagents, acids, suspended solids, sewage, etc., are generated in due course of mining. Huge amount, complex composition, wide range, and long duration are the features of wastewater emission that make it worth concern (Li et al. 2016).

Mining below the water table, i.e. underground mines and open pits, provide a direct conduit to aquifers. Infiltration of wastewater through overburden piles also potentially deteriorates groundwater (Thakur et al. 2013). The presence of hydraulic connections between surface water and groundwater is also a major source of contamination. During beneficiation, the chemicals used for leaching process dissolve metals and other toxic contaminants and form a highly acidic, metal-rich solution. This is the most potential source of water pollution as it leads to acid drainage and contaminant leaching in metal-mined areas. Since pH is the key factor in determining the leachability of heavy metals, a decrease in pH favors their mobility (Król et al. 2020). Leaching and blasting also result in raised concentrations of cyanide and nitrogen compounds in waters at mine sites. AMD or ARD occurs upon oxidation of metal sulfides and possesses long-term devastating impact on water bodies resulting in $pH < 4$, which challenges survival of flora and fauna. Natural weathering process also brings about oxidation of mineral metal sulfides leading to the formation of sulfuric acid, but that is a slow process. Mining and beneficiation expose sulfide-containing overburden to air and water and magnify the rate of chemical reactions leading to AMD (Geelani et al. 2013). It can also dissolve and discharge heavy metals into water bodies which are toxic, non-degradable, and persistent contaminants of environment having carcinogenic effects on living organisms as a result of bio-magnification. The effects of the presence of heavy metals in water bodies may be immediate (toxicity resulting in death) to sublethal, i.e. affecting growth and development. It is practically impossible to stop AMD after mining has begun for an acid-generating mine and to permit such project is least sustainable.

The mobility of pollutants from all the abovementioned sources is magnified in events of rainfall or snowfall, as these bring more water to mine site and contribute to runoff quantity, ultimately getting discharged and polluting water bodies (Geelani et al. 2013).

During the framing of mining project, the question that needs to be addressed is whether the water quality in mining-affected region will be fit for human consumption in future or whether it will be adequate to support native aquatic and terrestrial organisms. If not, then a treatment and recycle strategy must be designed based on the requirement. Water polluted due to mining may contain suspended solids, be highly saline, or acidic and/or contain special pollutants such as fluoride, heavy metals, and radioactive elements. Several purification techniques are, however, available and can be implemented for the treatment of mine water such as "high-efficiency clarification plus gravity valveless filter" for water containing suspended solids, reverse osmosis for highly saline mine water, limestone trench method or sulfate reduction bacteria reactor for acid mine water, ion exchange method for fluoride containing water and precipitation method for heavy metals (Wang et al. 2021). Treating and reusing mine water will not only lead to reduction in pollution due to discharge of wastewater but also avoid water shortage in future.

1.2.7 Sedimentation in Streams and Rivers

Both surface water and underground water are exploited during different phases of mining. In case of open-pit mining, a pit that extends up to below groundwater table is created and water is pumped out for mining to take place. This leads to the formation of pit lakes, and groundwater supply is turned off. On the other hand, placer mining pollutes surface waters by discharging large amounts of sediments into it. Sediment-laden surface runoff generally arises as sheet flow and gets collected in rills or gullies before being finally deposited in water bodies or floodplains (Geelani et al. 2013).

Additionally, disposal of tailings in streams and rivers is one of the worst practices done during mining operations. Mining waste deposits contain high amounts of potentially toxic elements and affect smaller stream sediments along with main streams close to their confluence. Toxic elements are then transported across long distances along major watercourse, thus resulting in regional pollution (Miler et al. 2022). A water management plan must be an indispensable part of any mining project as any meagre design may lead to high levels of dissolved as well as suspended solids in water resources nearby mine sites.

1.2.8 Groundwater Regime

The reaction of groundwater system to hydrological stresses and anthropogenic intervention is groundwater regime which is known to govern surface hydrology, soil texture, and vegetation (Basavaraja and Kumar 2020). Deep mine excavations under the groundwater table lower the groundwater level in adjacent areas. The water present is pulled by the action of gravity into the empty space and may lead to dewatering of wells, irrevocable loss of soil moisture and proves detrimental to revegetation during reclamation. Both quality and quantity of groundwater are affected adversely if the backfill material is significantly different from parent material or if a mine is situated in groundwater recharge zone.

1.2.9 Noise Pollution

Almost all the processes during mining involve use of heavy machinery during excavation, blasting, drilling, and transportation and are significant sources of noise pollution. Blasting operations are also the cause of vibrations experienced in areas near mine sites. Operation of heavy machinery and vehicles usually during overburden handling creates massive intensity of noise.

The World Health Organization (WHO) classifies noise above 65 dB as noise pollution. Noise above 75 dB is considered to be harmful and painful if it exceeds 120 dB. Noise level in mine sites often exceeds 90–120 dB, which is far above permissible limits. This may affect both physical and mental health such as damage hearing mechanism, severely affect memory, cause dizziness, insomnia, and fatigue, leading to disturbance among humans nearby and also affects wildlife distribution (Li et al. 2016).

1.2.10 Human Health

"Red Alert" – a documentary made by Saki (Non-governmental Organization) – highlights health problems faced by mine workers and shows that lung infections and heart problems are of high prevalence in mining areas (Basavaraja and Kumar 2020). Pollutants released from mine sites may enter humans from oral, dermal, or inhalation routes. The concentration of pollutants present in environment and its intake by organisms are taken into account to estimate the health risk. The amount of food ingested or the amount of air inhaled is useful for the calculation of daily consumption values, which is then compared with permissible limits (Pashkevich 2017). For substances like heavy metals, determining transfer effect (i.e. ratio of concentration in plants and soil) is important as plants are the basic component of food chain. Such pollutants tend to bio-accumulate while moving ahead in trophic levels. Heavy metals such as cadmium, zinc, and thallium are readily absorbed by the roots of plants, whereas arsenic, chromium, and mercury tend to settle on above ground parts. Various indices namely, estimated daily intake (EDI), target hazard quotient (THQ), hazard index (HI), and target cancer risk (TCR) have been stated by U.S. Environmental Protection Agency (US EPA) which are calculated using heavy metal concentration of edible portion of plants, reference dose (i.e. permissible limit), exposure duration, exposure frequency, average life span, and body weight of target population. They are used to estimate and highlight the risks associated with dietary exposure to potentially toxic elements over lifetime (Antoine et al. 2017).

1.2.11 Formation of Mine Dumps

The disposal of overburden leads to the formation of mine dumps, which transform the natural system to a technogenic landscape. Mine dump formation adversely affects the nature of underlying rocks along with modifying hydrological and hydrogeological regimes of surrounding area (Pashkevich and Petrova 2016). Additionally, it leads to the reduction of agricultural area, pollution and erosion of soil, and degradation of atmosphere in vicinity (Alekseenko and Pashkevich 2016). The volume and chemical properties of the wastes determine the magnitude of hazards associated with the dump site. Besides, the mobility and migration ability of wastes along with the degree of resistance provided by the environment are also pivotal in determining toxicity. Such anthropogenic alteration of soil and water composition in addition to air quality of an area modifies it to technogenic geochemical fluxes and halos (Pashkevich 2017).

On the basis of their genesis, mine dumps can be categorized into the following:

i) Piled dumps – consisting of structurally damaged soils, solid and mineral wastes of industrial origin, and solid wastes of domestic origin.
ii) Hydraulic-filled dumps – consisting of hydraulically stacked soils, household and mineral wastes (tailings and sludges).
iii) Technogenic sediments – consisting of solid mineral wastes redeposited through air and water (precipitated particles, technogenic sediments, and artificial soils). Sand, fly ash, salts, chemicals, and other inorganic compounds are examples of mineral pollutants, whereas materials of plant or animal origin along with resins, phenols, and alcohols constitute organic pollutants. Pathogenic microbes present in sewage water tending to enter water bodies are biological contaminants.

Table 1.2 Toxic pollutants released from mine dumps created by mining and processing industries.

Mine dump/industry	Pollutants released
Ferrous resources	Heavy metal(loid)s such as arsenic, cadmium, lead, mercury, chromium, nickel, zinc, and vanadium; acids and alkalis; and organic contaminants such as mineral oils
Non-ferrous resources	Heavy metal(loid)s such as arsenic, antimony, beryllium, cadmium, chromium, copper, lead, mercury, nickel, selenium, thallium, vanadium, and zinc; aromatic organic compounds such as cresol; and cyanides, fluorides, acids, and alkalis
Coal	Heavy metal(loid)s such as arsenic, chromium, lead, and zinc; organic compounds such as cresol, mineral oils, benzene, phenols, anthracene, toluene, xylene, and naphthalene; fluorides, acids, and bases
Petroleum	Heavy metal(loid)s such as arsenic, copper, chromium, lead, selenium, vanadium, and zinc; organic compounds such as anthracene, benzene, gasoline, mineral oil, naphthalene, phenol, and tar oil; acids and alkalis

The composition of mine dumps and range of pollutants they release vary with different mining industries. A list of potentially toxic pollutants has been enumerated in Table 1.2 (Pashkevich 2017).

Another factor that determines the degree of toxicity due to a mine dump is the migration of contaminants, which is controlled by a range of internal as well as external factors. External migration factors include temperature and pressure for atmosphere and lithosphere (Martínez-Sánchez et al. 2012); physicochemical conditions such as pH and Eh and parameters such as chemical bonds and gravitational properties are key determinants for hydrosphere (Chalov et al. 2016). Factors such as locomotion ability are crucial for biotic pollutants such as pathogenic bacteria. In case of atmospheric migration, pollutants migrate in a suspended state or in gaseous form and affect the receptors, i.e. plants, animals, and humans directly. In case of migration through water, both contaminated surface and groundwater merge into water bodies and reach receptors. But soils are generally considered to be indirect and intermediate mode of dissemination that particularly acts through food chains (soil–plants/animals–humans) (Pérez-Sirvent et al. 2016). A peculiar feature soils possess is their ability to accumulate, filter, neutralize, and decompose certain pollutants. Moreover, soils are naturally rich in microbes, support vegetation, and hence undergo remediation naturally.

1.2.12 Visual Impacts

Visual impacts are related to aesthetic aspects and rehabilitation of a mined site. Nearly all stages of mining from clearing of vegetation followed by removal of overburden and dumping deteriorate the aesthetics of a site. Rain splashes on overburden dumps lead to sheet, reel, or full erosion which further damages the aesthetic value of the site (Geelani et al. 2013). Mine dumps are known to affect terrain topography, geological structure, soil cover, hydrography, climate, and vegetation (Pashkevich 2017). Besides, disturbance in water drainage, water logging, soil erosion, destruction of vegetation, and dust blow are some direct impacts of mine dumps on human population residing nearby. The reduction in visual landscape value is characterized by declining landscape value. It is a parameter that cannot be measured quantitatively and is of subjective value (Geelani et al. 2013).

1.2.13 Social Impacts

While discussing the adverse effects of mining activities, its impact on environment is taken much into consideration; what is often taken for granted is the dislocation of forest dwellers and their lost home, land, and livelihood. The social impacts of mining are contentious. Though it can lead to creation of roads, employment of labor, development of inaccessible areas, and generation of wealth, the benefits are unevenly shared. Any commercial activity must ensure that the basic needs of dwellers are fulfilled and rights are not violated. This means they should get right to use land, get livelihood, and have access to clean water and safe place to dwell. They must be compensated fairly for the damages caused by mining to them. As per human laws, all citizens are equal and hence the marginal groups need to be protected. But in mining areas, the ones who are rich and powerful exploit the natural resources ruthlessly leading to depletion. The consequences are then faced by marginal dwellers whose livelihood relies on biomass-related products collected from immediate environment. Upon disappearance of forests and siltation of water bodies, their biomass-based subsistence economy is massively affected (Basavaraja and Kumar 2020). Even the common property resources such as community forests, village ponds, pastures, and wastelands, which must be available to all inhabitants, can be expected to be transferred under private ownership due to scarcity of resources as a result of rampant mining (Guha 2006).

1.2.14 Other Issues

Mining in India is disrespectful for human rights violation and adverse impacts on environment. Second, appropriate assessment of resources is not complete yet;

several areas still remain unexplored in terms of mineral presence. Since the distribution of minerals is uneven and fluctuates substantially from one area to another, absence of a genuinely evaluated data is a challenging issue.

Besides the abovementioned environmental problems, safety of mine workers and residents nearby also requires attention. Physical stability of mines is a crucial long-term concern keeping repercussions of slope failure in mind whereby toxic contaminants release and find a direct pathway to receptors (Mehta 2002). A pre-mining design with overburden disposal plan is a prerequisite of any mining project. An underestimation of the amount of overburden composing the slope may lead to failure. Besides slope failure, landslides in hilly terrains during mining operations are also not uncommon.

1.3 Economic Valuation of Impacts on Environment

Environment pays a huge cost for every mining project. These outcomes namely erosion, pollution, fertility loss, safety issues, and health risks should be estimated and evaluated with the production cost. It is necessary to accentuate the costs associated with the disposal of overburden which needs to be both physically and biologically stabilized. Assessment of abandonment cost of the mine inclusive of closure activities namely wire fencing surrounding work site, dismantling and demolition of structures, cleaning of site and machinery, remediation and reclamation, post-environmental monitoring and supervision charges, power cost, and miscellaneous costs needs to be done in order to ascertain economical valuation.

Moreover, conservation of resources and their administration should be done scientifically, usage of resources should be judicious, wastage should be minimal in extraction process, substitutes of the minerals being widely used should be sought out, logical recycling of used metals and scraps should be followed, and discovery of new deposits needs to be carried out in order to justify the costs associated with mining process.

1.4 Environment Protection and Policy Implication

The Government of India has framed various laws for environmental protection. But a strong legal system to check on their implementation is the need of hour. Both public and private sector industries visualize the economic benefits first, keeping environmental protection at stake. The judicial system needs to act as fast as possible in case of any illegal mining activity or infringement of environmental protection acts. Moreover, involving more and more people to spread awareness and providing safety to those who act as whistleblowers are needed to intensify environmental protection programs. No matter how big the company is, violators must be punished for illegal behavior and compulsorily engaged in rehabilitation of the site they harmed.

Permit to mining should be granted cautiously keeping environmental and biological fragility in mind. Areas that are rich in biodiversity and sensitive ecologically should be mapped and banned for mining. Wastelands generated during mining must be subjected to reclamation, and in order to ensure this, specialized bodies such as Reclamation trusts should be established to revive such lands and put them to use productively.

A well-formulated mining closure policy and its effective implementation are must for all nations. Further, government must necessitate eco-rating, i.e. obtaining certificates for mining companies on the basis of their environmental impacts. Post operation, mine owners need to certify that all protective measures were taken during mining and reclamation measures adopted were as per the approved plan. "Star Rating," an initiative by Ministry of Mines, Government of India, is awarded to mine leases in view of their endeavor to implement Sustainable Development Framework. This rating system will not only encourage and incentivize best practices but may also hasten clearances to be obtained from regulatory bodies in future.

1.5 Management and Reclamation

In order to restore environmental balance, active mining should have nominal effects on environment and abandoned sites must undergo reclamation. Reclamation, in general sense, means to repair disturbed or damaged site to its prior condition or for productive use. It is an umbrella term making use of restoration, rehabilitation, replacement, and remediation techniques either alone or together (Favas et al. 2018). While restoration aims to restore the mine site to its original conditions and is practically almost impossible, rehabilitation aims at returning to previous state partially. Rehabilitation emphasizes more on repairing ecosystem processes than ecosystem structure and function. Replacement is recreating a new site with economic and aesthetic value and has nothing to do with regeneration of original conditions (Favas et al. 2018). The adopted technique for reclamation is chosen on the basis of present conditions of mine site. It must be sustainable and eco-friendly besides being feasible to application. Remediation is a clean-up approach and specifically applied to contaminated sites before or along with reclamation techniques.

Since mine sites are generally located in remote areas, this "out-of-sight, out-of-mind" sort of negligence needs to be addressed. Various measures need to be adopted to manage and control abovementioned impacts. An integrated program encompassing several techniques needs to be implemented for reclamation process. The procedures can be delineated in an ordered process as follows:

a) **Preliminary assessment:** This is the first phase and involves interpretation of the present condition and diagnosis of environmental risk. It recognizes the source and composition of contaminants along with its dispersion process.
b) **Planned end state:** It is the conclusion of productive future use of the site (such as agricultural, forestry, nature conservation, landfill, and industrial) keeping environmental, technical, financial, and social facets in mind. This phase is crucial, since it influences all other phases of reclamation process in determining the methodologies adopted.
c) **Detailed assessment:** It involves interpreting nature and range of contaminants on the basis of preliminary assessment. A comprehensive analysis of samples (water, soil, and flora) is carried out, and toxicological profile of contaminants is studied along with their interactions with human and ecological receptors.
d) **Instituting clean-up criteria:** It is either based on regulatory limits, i.e. maximum permissible limits, or reflecting original site conditions, or compatible with planned end site use.
e) **Setting-up a remediation work plan:** Based on the regulations set by clean-up criteria, plan to control, treat, and dispose contaminated material is devised using physical (removal, incineration, and vitrification), chemical (oxidation and solvent extraction), and/or biological techniques (bioremediation and phytoremediation) (Venkateswarlu et al. 2016). Along with a remediation plan, a morphological recovery plan for scenic organization (backfilling and revegetation) of site is also designed.
f) **Final verification and monitoring of end site:** This is the last phase where the outcome of reclamation plan is evaluated. Soil and water quality are analyzed and compared with clean-up criteria to ensure their suitability with end site goals.

Historically, there have been several mine sites which have never been put through any reclamation program. These are examples of natural reclamation process brought about by local spontaneous vegetation. In places where technical intervention is not economically feasible, "natural recovery," "passive recovery," or "spontaneous succession" can be the most acceptable way for regeneration (Šebelíková et al. 2016). This process can be assisted anthropogenically to facilitate and accelerate vegetative colonization by wide-scale planning of natives, seed dispersal, and addition of organic amendments (Chibrik et al. 2016).

Besides the abovementioned reclamation strategy, a lot can be done during mining sessions. Precautions should be taken during overburden disposal. During the planning phase itself, a site with solid, impermeable base should be chosen for overburden disposal in order to ensure minimum leaching during events of precipitation. It should be located far away from water bodies, if not, then water streams need to be diverted to avoid any contact.

Before the implementation of environmental laws, most mining projects dumped the tailings in the nearest location possible, which is now universally disapproved. The most preferred method is to dewater and dispose dry tailings as backfill. The use of tailings pond or wet tailing impoundment facility is by far the most common. If the mine site is located in close proximity to sea, submarine tailings disposal is proposed (Environmental Law Alliance Worldwide 2010).

Overburden dumps should be made after studying topography of the area and determining wind flow pattern, so as to minimize the contact between overburden and wind. The height of overburden piles should be reduced in order to minimize wind erosion. Water sprays can be used to control dust particle generation during mining operations and overburden disposal. Overburden and mine waste can also be utilized for backfilling the terrain in order to restore it. It can be accompanied by surface levelling and providing a soil cover for vegetation. The waste can be crushed to optimum size and used as road metal later. After mining closure, vegetative covers can control erosion helping in stabilization and reclamation of a mine site.

For controlling air pollution, clean energy sources like compressed natural gas (CNG) and liquefied petroleum gas (LPG) should be used as fuels for machine equipments and transportation activities during mining. In order to control noise pollution, the slope or gradient of the haul road should be optimized as much as possible. Since the noise generated is directly proportional to the power required by haul trucks' engine, noise can only be minimized by reducing the gradient of haul road (Geelani et al. 2013).

Revegetation is by far the most preferred and common approach applied for reclamation of mine sites. Establishing vegetative cover over the entire site or in selective patches can be considered as per the intended use. But the physicochemical conditions of soil tend to retard plant growth and establishment. Displacement of topsoil, drought, erosion, compaction, presence of heavy metals, and inadequacy of nutrients and beneficial microbes are some characteristics of mine sites that hinder revegetation and hence reclamation (Favas et al. 2016). Ecological restoration of mine land requires rigorous knowledge about relationship between edaphic factors and vegetation, out of which soil texture is the most important factor for the development of vegetation. A case study from Bor mines identified hemicryptophytes of psammophytic character to be the most successful primary colonizers during human-assisted natural recovery, suggesting assessment of edaphic factors to be essentially done prior to reclamation project (Ranđelović et al. 2014).

Moreover, human intervention is also needed for amendment of soil with organic fertilizers such as compost, manure, bio-solids, and treated sewage sludge in order to restore fertility (Kidd et al. 2015; Favas et al. 2016). This is also a feasible way to recycle organic wastes in the most sustainable manner. Considering the nature of soil and harsh conditions of environment, drought-resistant varieties of fast growing plants capable of growing in nutrient-deficient metalliferous soils can be selected for possible plantation. This plant cover would not only be able to tolerate hostile soil but also control erosion of soil and leaching of contaminants to lower levels (Kopittke et al. 2012). Succession during establishing plant cover, i.e. initially herbs and eventually shrubs and trees, will check the dissemination of pollutants and minimize permeation of rain water. Roots of plants will also promote soil aeration along with performing phytoremediation (Prasad 2015).

A wider outlook in this approach requires application of specific plants to solve particular purpose. Hyperaccumulator plants can be applied for phytoextraction purpose since they have the ability to accumulate heavy metals in harvestable parts, with caution exercised on their disposal as they pose a risk of introducing toxic elements in food chain (Venkateswarlu et al. 2016). In order to reduce mobility and bioavailability of contaminants, phytostabilization technique using suitable plant and associated microbial species can be applied, which not only attains pollution control but also minimizes the threats to herbivores and humans (Sheoran et al. 2010). In this context, *Agrostis stolonifera*, a grass species, has been identified as a potential phytostabilizer in copper tailings (Kasowska et al. 2018). Plant inventories can be created for the identification of native species in mine areas with phytostabilization potential (Santos et al. 2017).

Leguminous plants can be employed to enrich soil with nitrogen; mycorrhizal fungi can be inoculated to increase phosphorous bioavailability; planting annuals with perennials will constantly enrich soil with organic matter, each of these measures can be synchronized for their distinct and cumulative aid in soil fertility (Sheoran et al. 2010).

Hence revegetation of abandoned mine site can lead to stabilization and attenuation of pollutants, in addition to being visually pleasing. Advanced engineering technologies can be synchronized with vegetation covers to serve purposes to enclose or encapsulate toxic wastes generated from mining (Favas et al. 2016). Engineering techniques, namely biological greening bags, eco-stick slope protection, container seedling planting, and vegetative-carpet greening measure, can also be applied to hasten the process of vegetation recovery (Zhang et al. 2014).

1.6 Progressive Reclamation

An ideal mining closure plan consists of two components; first, a progressive or concurrent plan that conducts activities being done along with mining and second, a final plan that begins after mining (Tiwari and Chakraborty 2001; Kabir et al. 2015). Progressive reclamation works out post closure activities concurrently with mining activities. It involves integrated mine planning for closure and reclamation, environmental studies, and financial funding even before actual mining process starts. During mining process, it plans for mitigating climate change impacts, water management, and land use. Reclamation of patches not under use can be done while other patches are being mined in a progressive plan. After mining closure, the evidence of mining needs to be removed as much as possible, closure and reclamation activities need to take local environmental conditions into account, and evaluation of rehabilitation and reuse of land is done.

Progressive recovery may include removing, storing, and using overburden for reclamation; implementing research results to achieve ideal remediation; and establishing favorable nurse crops for revegetation (Favas et al. 2018). To summarize, progressive mining involves continuous monitoring throughout the life cycle of mine; continuous engagement and discussions with native people, communities, and regulators; continuous updates to ascertain closure plans complement any alterations to the mine during operations. It ensures that rehabilitated landscape can support life for upcoming centuries.

1.7 Conclusion

Consequently, there are many evident as well as not so evident effects of mining on environment. But issues need to be addressed aptly for sustainable development. The mining activities within a nation should be closely monitored and coordinated with environmental protection and preservation acts. An abiding and practicable reclamation strategy requires interdisciplinary expertise and involves alliance among pedologists, chemists, engineers, botanists, ecologists, landscapers, etc., to make reclamation successful. Even though the foremost requirement of mining industry is to carry out mining judiciously and implementing an effective mining closure policy so as to reduce the number of abandoned mine lands, reclamation of existing mine lands should also be taken into account.

References

Alekseenko, A.V. and Pashkevich, M.A. (2016). Novorossiysk agglomeration landscapes and cement production: geochemical impact assessment. *IOP Conference Series: Earth and Environmental Science*, Tomsk, Russia (4–8 April 2016). p. 12050.

Antoine, J.M.R., Fung, L.A.H., and Grant, C.N. (2017). Assessment of the potential health risks associated with the aluminium, arsenic, cadmium and lead content in selected fruits and vegetables grown in Jamaica. *Toxicology Reports* 4: 181–187.

Barve, A. and Muduli, K. (2011). Challenges to environmental management practices in Indian Mining industries. *International Conference on Innovation, Management and Service IPEDR*, vol. 14. Singapore: IACSIT Press.

Basavaraja, T. and Kumar, V. (2020). Health conditions of mining labourers in Karnataka. *International Journal of Creative Research Thoughts* 8 (3).

Buta, M., Blaga, G., Paulette, L. et al. (2019). Soil reclamation of abandoned mine lands by revegetation in Northwestern part of Transylvania: a 40-year retrospective study. *Sustainability* 11: 3393. https://doi.org/10.3390/su11123393.

Chalov, S., Thorslund, J., Kasimov, N. et al. (2016). The Selenga River delta: a geochemical barrier protecting Lake Baikal waters. *Regional Environmental Change* 1–15.

Chibrik, T.S., Lukina, N.V., Filimonova, E.I. et al. (2016). Biological recultivation of mine industry deserts: facilitating the formation of phytocoenosis in the Middle Ural region, Russia. In: *Bioremediation and Bioeconomy* (ed. M.N.V. Prasad), 389–418. USA: Elsevier.

Corporate Catalyst India (CCI). (2010). Mineral and mining industry in India. http://www.cci.in/pdf/surveys_reports/mineral-mining-industry.pdf (accessed 6 December 2021).

DeMeo, D.L., Zanobetti, A., Litonjua, A.A. et al. (2004). Ambient air pollution and oxygen saturation. *American Journal of Respiratory and Critical Care Medicine* 170: 383–387.

Emmanuel, A.Y., Jerry, C.S., and Dzigbodi, D.A. (2018). Review of environmental and health impacts of mining in Ghana. *Journal of Health Pollution* 17: 43–52.

Environmental Law Alliance Worldwide (2010). *Guidebook for Evaluating Mining Project EIAs*. Environmental Law Alliance Worldwide. https://www.elaw.org/mining-eia-guidebook (accessed 30 November 2021).

Favas, P.J.C., Pratas, J., Chaturvedi, R. et al. (2016). Tree crops on abandoned mines for environmental remediation and industrial feedstock. In: *Bioremediation and Bioeconomy* (ed. M.N.V. Prasad), 219–249. Amsterdam: Elsevier.

Favas, P.J.C., Martino, L.E., and Prasad, M.N.V. (2018). Abandoned mine land reclamation—challenges and opportunities (holistic approach). In: *Bio-Geotechnologies for Mine Site Rehabilitation*, 3–31. Elsevier. http://doi.org/10.1016/B978-0-12-812986-9.00001-4.

Franks, D.M., Cohen, T., McLellan, B., and Brereton, D. (2010). Technology Futures Discussion Paper: Technology Assessment and the CSIRO Minerals Down Under National Research Flagship. Prepared for CSIRO Minerals Down Under Flagship, Minerals Futures Cluster Collaboration, July, Centre for Social Responsibility in Mining, Sustainable Minerals Institute, The University of Queensland, Brisbane.

Geelani, S.M., Bhat, S.J.A., Geelani, S.H. et al. (2013). Mining and its impacts on environment with special reference to India. *International Journal of Current Research* 5 (12): 3586–3590.

González-Valoys, A.C., Arrocha, J., Monteza-Destro, T. et al. (2022). Environmental challenges related to cyanidation in Central American gold mining; the Remance mine (Panama). *Journal of Environmental Management* 302 (A): 113979.

Guha, R. (ed.) (2006). *Social Ecology*. New Delhi: Oxford University Press.
Kabir, S.M.Z., Rabbi, F., Chowdhury, M.B. et al. (2015). A review of mine closure planning and practice in Canada and Australia. *World Review of Business Research* 5 (3): 140–159.
Kasowska, D., Gediga, K., and Spiak, Z. (2018). Heavy metal and nutrient uptake in plants colonizingpost-flotation copper tailings. *Environmental Science and Pollution Research* 25: 824–835. https://doi.org/10.1007/s11356-017-0451-y.
Kidd, P., Mench, M., Álvarez-López, V. et al. (2015). Agronomic practices for improving gentle remediation of trace element-contaminated soils. *International Journal of Phytoremediation* 17: 1005–1037.
Kopittke, P.M., Blamey, F.P.C., and Menzies, N.W. (2012). Guidance document for the revegetation of land contaminated by metal(loid)s, Adelaide. CRC CARE Technical Report No. 20, CRC for Contamination Assessment and Remediation of the Environment.
Krishnamurthy, S.V. (2003). Amphibian assemblages in undisturbed and disturbed areas of Kudremukh National Park, central Western Ghats, India. *Environmental Conservation* 30 (03): 274–282.
Król, A., Mizerna, K., and Bożym, M. (2020). An assessment of pH-dependent release and mobility of heavy metals from metallurgical slag. *Journal of Hazardous Materials* 384: 121502.
Kumar, S.V. (2019). *Exploration and Mining in India: Time for a Deeper Look*. TERI Discussion Paper. New Delhi: The Energy and Resources Institute.
Li, J., Zhang, T.T., Yang, W., and Zhang, Y. (2016). The environmental impact of mining and its countermeasures. *MATEC Web of Conferences* 63: 04010.
Martínez-Sánchez, M.J., García-Lorenzo, M.L., Pérez-Sirvent, C. et al. (2012). Trace element accumulation in plants from an aridic area affected by mining activities. *Journal of Geochemical Exploration* 123: 8–12.
Mehta, P.S. (2002). The Indian mining sector: effects on the Environment and FDI inflows. *CCNM Global Forum on International Investment*, Paris, France (7–8 February 2002).
Miler, M., Bavec, S., and Gosar, M. (2022). The environmental impact of historical Pb-Zn mining waste deposits in Slovenia. *Journal of Environmental Management* 308: 114580.
Pashkevich, M.A. (2017). Classification and environmental impact of mine dumps. In: *Assessment, Restoration and Reclamation of Mining Influenced Soils* (ed. J. Bech, C. Bini and M.A. Pashkevich), 1–32. Elsevier.
Pashkevich, M.A. and Petrova, T.A. (2015). The investigation of toxic waste burial polygon impact on the atmospheric air using lichenoindication method (Russia). *International Journal of Ecology and Development* 30 (3): 98–105.
Pashkevich, M.A. and Petrova, T.A. (2016). Technology of artificial deposit preservation at the Kursk magnetic anomaly. *Ecological Engineering* 17 (4): 11–16.

Pérez-Sirvent, C., Hernández-Pérez, C., Martínez-Sánchez, M.J. et al. (2016). Geochemical characterisation of surface waters, topsoils and efflorescences in a historic metal-mining area in Spain. *Journal of Soils and Sediments* 16 (4): 1238–1252.

Prasad, M.N.V. (2015). Engineered phyto-covers as natural caps for containment of hazardous mine and municipal solid waste dump sites: possible energy sources. In: *Phytoremediation for Green Energy* (ed. M. Öztürk, M. Ashraf, A. Aksoy and M.S.A. Ahmad), 55–68. Dordrecht, Heidelberg, New York, London: Springer.

Ranđelović, D., Cvetković, V., Mihailović, N., and Jovanović, S. (2014). Relation between Edaphic factors and vegetation development on copper mine wastes: a case study from Bor (Serbia, SE Europe). *Environmental Management* 53 (4): 800–812.

Santos, A.E., Cruz-Ortega, R., Meza-Figueroa, D. et al. (2017). Plants from the abandoned Nacozari mine tailings: evaluation of their phytostabilization potential. *PeerJ* 5: e3280. https://doi.org/10.7717/peerj.3280.

Šebelíková, L., Řehounková, K., and Prach, K. (2016). Spontaneous revegetation vs. forestry reclamation in post-mining sand pits. *Environmental Science and Pollution Research* 23: 13598–13605.

Sheoran, V., Sheoran, A.S., and Poonia, P. (2010). Soil reclamation of abandoned mine land by revegetation: a review. *International Journal of Soil, Sediment, and Water* 3 (2): 13.

Tiwari, R. and Chakraborty, M.K. (2001). A progressive mine closure plan for a semi-mechanized quarry of dolomite. *Indian Mining & Engineering Journal* 50 (2): 31–36.

Thakur, B.K., Gupta, V., and Chattopadhyay, U. (2013). Arsenic groundwater contamination related socio-economic problems in India: issues and challenges. In: *Knowledge Systems of Societies for Adaptation and Mitigation of Impacts of Climate Change Environmental Science and Engineering*, 163–182. Springer.

Varshney, C.K., Garg, J.K., Lauenroth, W.K., and Heitschmidt, R.K. (1979). Plant responses to sulfur dioxide pollution. *C R C Critical Reviews in Environmental Control* 9 (1): 27–49. https://doi.org/10.1080/10643387909381667.

Venkateswarlu, K., Nirola, R., Kuppusamy, S. et al. (2016). Abandoned metalliferous mines: ecological impacts and potential approaches for reclamation. *Reviews in Environmental Science and Biotechnology* 15 (2): 327–354.

Wang, X., Gao, Y., Jiang, X. et al. (2021). Analysis on the characteristics of water pollution caused by underground mining and research progress of treatment technology. *Advances in Civil Engineering* 9984147.

Yusuf, M. and Arisoesilaningsih, E. (2017). Exotic plant species attack revegetation plants in post-coal mining areas. *AIP Conference Proceedings* 1908: 040002.

Zhang, Y., Yang, J., and Wu, H. (2014). Dynamic changes in soil and vegetation during varying ecological-recovery conditions of abandoned mines in Beijing. *Ecological Engineering* 73: 676–683.

Zipper, C., Angel, P., and Adams, M.B. (2019). The Forestry Reclamation Approach: An Essential Tool for Controlling Invasive Exotic Plants on Active Mine Sites, Forest Reclamation Advisory No. 17.

2

Soil Contamination, Risk Assessment, and Phytoremediation of Mine Land

Abstract

Mining activities have led to the formation of dumps and abandoned mine lands which have negative impacts on environment immediately and even after several years. Mine lands pose environmental risk due to the presence of moving and transforming heavy metals. Several risk assessment methods via calculation of pollution indices have been designed to perform an extensive geochemical assessment to analyze soil quality of mine lands, to carry out ecological as well as human health risk assessment, to determine the origin of pollutants (natural or anthropogenic), and to study the degree of soil degradation. Reversing soil degradation, regaining soil fertility, and remediation of polluted soil are the major challenges and need redressal immediately after mining operations. Revegetation emerges out to be a feasible solution for ecosystem restoration.

Keywords *Mine spoil land; heavy metals; toxicity; pollution index*

CONTENTS

Eco-Restoration of Mine Land, First Edition. Vimal Chandra Pandey, Jitendra Ahirwal, Roopali Roychowdhury, and Ritu Chaturvedi.

2.1 Introduction

Soil is a complex and dynamic system witnessing a range of physical, chemical, biological, geological, and bio-geo-physicochemical reactions over time and space (Tamene and Seyoum 2015). Soil biodiversity, health, and resilience are crucially restricted in harsh environments along with being highly sensitive to human activities (Fazekašová and Fazekaš 2020). Heavy metal accumulation in soil can be a result of both natural and anthropogenic causes. Since heavy metals are notable components of the Earth's crust (Zhou et al. 2015), weathering and pedogenesis processes at a site can lead to conditions conducive for metal accumulation (Kierczak et al. 2016). The extraction of minerals from ores by humans has led to the mobilization and release of heavy metals into the environment. Mining and processing industries are one of the most environmentally perturbing industries, but indispensable for the economy as minerals constitute around >70% of resources (by volume) globally (Pashkevich 2017). In India, more than 0.8 million hectares of land are under mining, of which a considerable area is constituted by forest lands. India inhabits around 20 000 known deposits of minerals (11 – metallic, 52 – nonmetallic, 4 – fuels, and 22 – minor minerals) (Geelani et al. 2013).

Mining activities have adverse effects on environment at local, regional, and global scales (e.g. emission of greenhouse gases during extraction and transportation operations). Nonetheless, the adversity of any mining operation is acutely felt at local level. Soil degradation, soil pollution, seismic changes, air pollution through dust and emissions, surface and groundwater pollution, increased noise levels, loss of biodiversity, etc., are some adversities faced at local scale. Figure 2.1 presents the adverse effects of mining at physical, chemical, biological, and aesthetical levels. The release of heavy metals into soil is a serious outcome of mining and treatment of ores (Fazekaš 2018). The increasing concentration of heavy metals, especially the mobile ones, is not just hazardous for soil and water quality but also poses a risk for human health at the same time (Yang et al. 2018). With intensive industrial revolution including mining operations, which is a principal cause of heavy metal contamination, evaluation of ecological risk becomes necessary (Kowalska et al. 2016). Studies pertaining to heavy metal analysis of soil samples

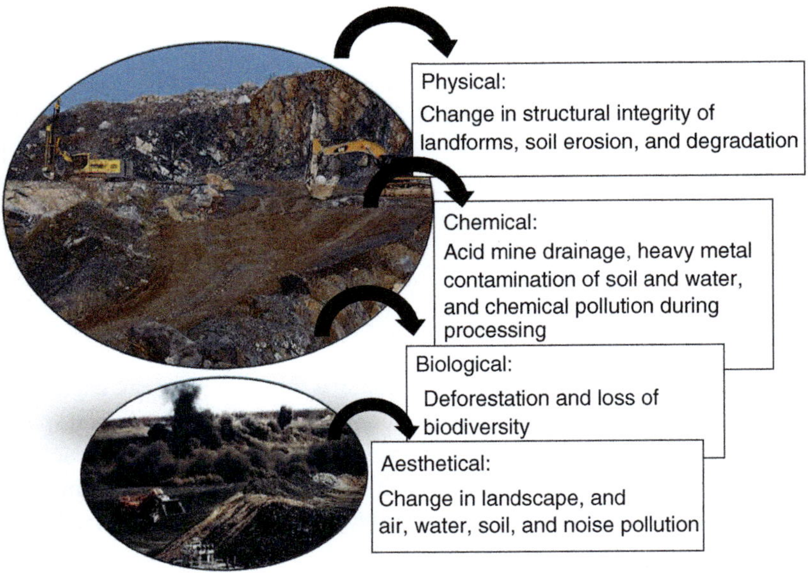

Figure 2.1 Effects of mining on environment.

in chronological order clearly show that there is a perpetual likelihood to increased accumulation as a result of rapid but unsustainable development (Hu et al. 2013). Such information about progressive soil degradation necessitates appropriate assessment using precise tools and its management (Gong et al. 2008).

The assessment of risk to ecological and human health is generally done by analyzing soil samples from study sites systematically and orderly, followed by comparison with threshold or allowable values (Nedyalkova and Simeonov 2019).

The data presented in several geochemical studies have come up with the formation of an exhaustive database of heavy metal concentrations in soil which can be applied as background values for ecological studies (Obiora et al. 2016). However, comparing the present-day values with these background values can only estimate the probability of contamination without determining overall soil quality (Caeiro et al. 2005). Moreover, application of computational tools to derive statistical relationships (e.g. correlation) between soil properties and heavy metal concentrations is not sufficient to assess soil contamination (Kowalska et al. 2016). A comprehensive and qualitative risk assessment tool is an empirical requisite for estimating the hazards to human health presented by toxic elements through various routes of exposure, i.e. inhalation, ingestion, and dermal absorption (skin contact) (Agyeman et al. 2021). Therefore, in order to make an accurate evaluation of soil contamination, pollution indices can be employed for carrying geochemical assessment in a comprehensive manner. Besides estimating environmental risk and degree of soil degradation (Caeiro et al. 2005), these indices can also predict the cause of soil contamination being natural or anthropogenic (Gong et al. 2008) along with surveilling the quality of soil and safeguarding sustainability for future, chiefly in the case of agro-ecosystems (Kelepertzis 2014).

Upon assessing the hazards associated with the presence of toxic elements, a plan for remediation is designed not only to prevent further spread to receptors but also to restore the mine site as best as possible. In order to remediate mine tailings and abandoned mine lands, ecological engineering needs should be practiced and a feasible remediation option should be employed. Physical and chemical methods namely excavation, soil washing, replacement, immobilization, and vitrification besides being labor extensive and expensive also sometimes require movement of polluted soil to remediation sites. In situ methods offer the advantages of being cost effective and eco-friendly, and phytoremediation is one such suitable method. Phytoremediation involves use of physiological processes of living plants to extract or immobilize contaminants from contaminated matrix. A variety of terrestrial and aquatic plant species can be used for remediation of soil and water polluted with a wide-ranging organic and inorganic contaminants like metals (Cr, Cd, Hg, and Pb), metalloids (As and Sb), radioactive elements (U and Sr), herbicides (atrazine, nitroaromatic compounds, and chlorinated compounds), pesticides, petroleum hydrocarbons (BTEX), explosives (TNT), and industrial organic wastes (PCPs, PCBs and PAHs) (Ensley 2000). Phytoremediation has been extensively employed in remediation and restoration of abandoned mine lands (Favas et al. 2014; Buta et al. 2019).

Further, in order to estimate the level of soil degradation as a result of mining project and also to determine the progress of reclamation project, risk assessment indices can be employed to analyze not only pre-remediation and post-remediation status of mine land but also ongoing remediation status.

2.2 Soil Contamination

Investigation of soil profile for its heavy metals can give insights about its origin and historical anthropogenic activities across the vicinity (Sołek-Podwika et al. 2016). Areas undergoing mining and associated activities witness high concentrations of toxic elements in soil and water bodies (both surface and underground), which represent geochemical risks to humans and environment (Mileusnić et al. 2014). The tailings produced as a result of ore processing contain high levels of toxic metals which constantly pollutes environment due to erosion and leaching (Baragaño et al. 2019). Acid mine drainage (AMD), also referred to as acid rock drainage (ARD), is one of the most hazardous consequences associated with mining operations. It generally occurs if the mined material comes in contact to oxygen and water. It is characteristically rich in concentrations of heavy metals (As, Cd, Cu, b, Zn, etc.) and possesses low pH, capable of severely polluting soil and water bodies (Ma et al. 2019). The problem of heavy metal contamination of soil and water poses a threat to food security globally (Chen et al. 2015); its incorporation in trophic chain will eventually lead to health hazards. Acute exposure to arsenic, a carcinogenic heavy metal, can lead to hypertension, cardiovascular, and gastrointestinal disorders (Jomova et al. 2011). Lead can induce anemia, encephalopathy, impair kidney functions, and cause a range of neurological problems. Cadmium is primarily toxic to kidneys and skeletal and respiratory systems as well. It is also a human carcinogen. The persistence of heavy metal contamination of soil is mainly due to soil's geo- and bioaccumulation potential in addition to its transport kinetics within soil profile (Fagbot and Olanipekun 2010; Kowalska et al. 2018).

Mining projects differ according to the type of mineral being extracted from the Earth. But in almost all cases, metallic ores are found deep buried under layers of ordinary soil or rocks (referred to as waste rock or overburden) that need to be removed to successfully extract the ore. Hence, excavation of overburden is technically the first step that makes difference in different mining projects. During open-pit mining (i.e. a strip mining, when the ore exists very deep underground), removal of vegetation and deforestation precedes the extraction process, leading to loss of biodiversity. Thereafter, overburden is removed via heavy machinery, whereby crushing of rocks generates dust and leads to air pollution. While during the separation process, liquids from rock slurries (containing heavy metals) may leak into bedrocks (Kusin et al. 2018). Therefore, it is one of the most environmentally destructive types of mining processes. If the ore of interest is associated with sediment in a streambed or floodplain, hydraulic mining is performed. Since it takes place within a streambed, it leads to severe pollution of water bodies. In case of underground mining, certain amount of overburden needs removal to grant access to ore deposit and thereafter a network of tunnels or shafts is stretched to directly access the ore. This kind of mining is costlier and entails safety risks for miners due to explosions. It leads to loss of structural integrity of rocks. Whatever be the method opted by a mining project, an enormous quantity of overburden is produced. Strip ratio is the ratio between overburden and the quantity of mineral ore, is usually greater than one, and can be much higher. This high volume waste containing toxic elements is generally stockpiled or dumped within underground mine and is hazardous to the environment. The next step of mining is grinding the ore and separating the metal of interest in a process called "beneficiation." This process again leads to generation of huge quantities of wastes, referred to as tailings. Some projects also demand re-extraction of metals from tailings, if it is economically beneficial. This can be performed on-site or off-site as well, but still entails their environmental impact during waste disposal process. The success of a mining project depends on the waste disposal technology adopted and determines its environmental acceptability. The key long-term goal of any mining project should be prevention of mobilization of toxic elements from the tailings. A mine closure plan must also address its policy and measures undertaken to restore the site that utmost resembles its pre-mined quality.

2.3 Risk Assessment

Risk, as per Environmental Protection Agency (EPA), can be considered as the chances of harmful effects to ecological systems or human health due to exposure to an environmental stressor. Stressors are physical/chemical/biological entities capable of affecting natural resources or entire ecosystems and inducing adverse response. EPA assesses risk not only to identify the type and extent of health risks to humans and ecological receptors (fauna) from stressors (i.e. pollutants, etc.) present in the environment but also to predict the behavior of pollutants in future. Risk assessment usually begins with a planning stage where the motive and scope of assessment are ascertained. Thereafter, nature and magnitude of contaminants are determined along with predicting their fate and impact in future. This is followed by evaluation of exposure frequency and inherent toxicity to predict the likelihood and extent of adverse health response in future.

Environmental risk can be assessed at two levels: ecological and human health.

2.3.1 Ecological Risk Assessment

It is the evaluation of impact of contaminant exposure on environment. In order to carry out ecological risk assessment, the first and foremost step is "Problem Formulation," i.e. to identify the ecological entity (species, community, ecosystem, habitat, or biome) to be protected and its specific attributes facing threat. The next step is "Analysis," i.e. determination of level of exposure of entities to stressors. Here calculations are made regarding bioavailability and bioaccumulation of contaminant to plants and animals. The last step is "Risk Characterization" where risk is estimated by studying exposure profile and effects and described to provide information regarding harmful effects on entities of concern.

A critical assessment of soil contaminated with heavy metals can be performed using pollution indices in order to make a comprehensive account of the degree of contamination. Pollution indices are useful not only to evaluate soil quality but also to predict sustainability of ecosystem (Kowalska et al. 2018). These indices act as indicators to test the presence and intensity of soil contamination and are a channel for processing, analyzing, and conveying raw environmental information to environmentalists, researchers, decision makers, and public (Caeiro et al. 2005).

They can be classified into (i) single or individual and (ii) complex or integrated based on the number of pollutants being analyzed.

Single indices group, e.g. single pollution index (PI), threshold pollution index (PI_T), contamination factor (Cf), enrichment factor (EF), single index of ecological risk factor (E_r^i), biogeochemical index (BGI), and geoaccumulation index (I_{geo}), contain instruments which are implemented for unitary assessment of soil contamination, i.e. calculated for each heavy metal individually. It requires the knowledge of geochemical background values or any other references such as preindustrial concentrations, threshold level, and average crust level, in addition to the heavy metal concentration in soil under study. These values can be categorized into grades as shown in Table 2.1.

Complex indices, e.g. sum of contamination (PI_{sum}), average single pollution index (PI_{avg}), vector modulus of pollution index (PI_{Vector}), weighted average of pollution index (PIw_{Avg}), background enrichment factor (PIN), integrated pollution index (IPI), degree of contamination (Cdeg), modified degree of contamination (mCd), pollution load index (PLI), potential ecological risk (PERI), Nemerow pollution index (P_N), contamination severity index (CSI), multi-element contamination (MEC), the probability of toxicity (MERMQ), and exposure factor (ExF), are holistic as they consider more than one heavy metal or the sum of their individual indices and allow the specification of degree of contamination comprehensively. They require individual as well as total concentration of toxic elements in soil analyzed. The values obtained can be used to estimate the soil quality as shown in Table 2.2.

Background Values: A key step for analysis of soil contamination is to establish a reference or geochemical background for comparison which is a baseline sample of known metal concentration. It demarcates normal concentrations of heavy metal in soil from abnormal ones. They should not be higher than the allowable or threshold value, i.e. the upper limit of concentration of normal heavy metal in soil (Reimann and Garret 2005). For this purpose, either average crustal value can be used for reference or a baseline level can be computed by analyzing local soil unaffected from anthropogenic activities. Since average crustal values do not consider natural geochemical variability and are derived using bulk fraction, it would be unfair to compare them with fine fraction soil concentrations. In order to make an appropriate comparison, establishment of reference values using the soil that is similar in origin is more accepted approach. To accomplish this practice, soil collected from upper layer can be compared to samples taken from deeper layers across the same core. The selection of less contaminated samples for setting up baseline average values is most reliable when metal trend is investigated in lower core (Abrahim and Parker 2008) as deeper layers are least impacted in due course of development and reflect preindustrial concentration of above layers.

Table 2.1 Classes of single indices to estimate soil quality.

Single indices	Value	Contamination level
Single pollution index (PI)	$PI \leq 1$	Nonpolluted
	$1 < PI \leq 2$	Mildly polluted
	$2 < PI \leq 3$	Moderately polluted
	$PI > 3$	Severely polluted
Threshold pollution index (PI_T)	$PI_T < 1$	Unpolluted
	$1 < PI_T \leq 2$	Low polluted
	$2 < PI_T \leq 3$	Moderately polluted
	$3 < PI_T \leq 5$	Strongly polluted
	$5 < PI_T$	Very strongly polluted
Contamination factor (Cf)	$Cf < 1$	Low
	$1 < Cf < 3$	Moderate
	$3 < Cf < 6$	Considerable
	$Cf > 6$	Very high
Enrichment factor (EF)	$EF < 2$	Deficiency to minimal enrichment
	$2 < EF > 5$	Moderate enrichment
	$5 < EF > 20$	Significant enrichment
	$20 < EF > 40$	Very high enrichment
	$EF < 40$	Extremely high enrichment
Ecological risk factor (E_r)	$E_r < 40$	Low
	$40 < E_r > 80$	Moderate
	$80 < E_r > 160$	Considerable
	$160 < E_r > 320$	High
	$320 < E_r$	Very high
Biogeochemical index (BGI)	$BGI < 1$	Low
	$1 < BGI < 5$	Apparent
	$5 < BGI < 10$	Moderate
	$10 < BGI < 15$	Considerable
	$15 < BGI < 20$	High
	$BGI > 20$	Very high
Geoaccumulation index (I_{geo})	$I_{geo} \leq 0$	Uncontaminated
	$0 < I_{geo} < 1$	Uncontaminated to moderately contaminated
	$1 < I_{geo} < 2$	Moderately contaminated
	$2 < I_{geo} < 3$	Moderately to heavily contaminated
	$3 < I_{geo} < 4$	Heavily contaminated
	$4 < I_{geo} < 5$	Heavily to extremely contaminated
	$I_{geo} \geq 5$	Extremely contaminated

Table 2.2 Classes of complex indices to study soil quality.

Complex indices	Value	Contamination level
New pollution index (PIN)	0–7	Clean (Class 1)
	7–95.1	Traces of contamination (Class 2)
	95.1–518.1	Lightly contaminated (Class 3)
	518.1–2548.6	Contaminated (Class 4)
	>2548.6	Highly contaminated (Class 5)
Integrated pollution index (IPI)	≤ 1	Low level of pollution
	$1 < \text{IPI} \leq 2$	Moderate level of pollution
	$2 < \text{IPI} \leq 5$	High level of pollution
	>5	Extremely high
Contamination degree (Cdeg)	≤ 8	Low contamination
	$8 < \text{Cdeg} \leq 16$	Moderate contamination
	$16 < \text{Cdeg} \leq 32$	Considerable contamination
	$32 < \text{deg}$	Very high contamination
Modified contamination degree (mCd)	$\text{mCd} < 1.5$	Nil to very low degree of contamination
	$1.5 \leq \text{mCd} < 2$	Low degree of contamination
	$2 \leq \text{mCd} < 4$	Moderate degree of contamination
	$4 \leq \text{mCd} < 8$	High degree of contamination
	$8 \leq \text{mCd} < 16$	Very high degree of contamination
	$16 \leq \text{mCd} < 32$	Extremely high degree of contamination
	$\text{mCd} \geq 32$	Ultra-high degree of contamination
Pollution load index (PLI)	$\text{PLI} < 1$	Perfect
	$\text{PLI} = 1$	Only baseline levels of pollutants are present
	$\text{PLI} > 1$	Deterioration of soil quality
Potential ecological risk index (PERI)	$\text{RI} < 150$	Low
	$150 < \text{RI} > 300$	Moderate
	$300 < \text{RI} > 600$	Considerable
	$600 \leq \text{RI}$	Very high

Table 2.2 (Continued)

Complex indices	Value	Contamination level
Nemerow pollution index (P_N)	$P_N \leq 0.7$	Clean
	$0.7 < P_N \leq 1$	Warning limit
	$1 < P_N \leq 2$	Slight pollution
	$2 < P_N \leq 3$	Moderate pollution
	$P_N > 3$	High pollution
Contamination severity index (CSI)	≤0.5	Uncontaminated
	0.5 < CSI ≤ 1	Very low severity
	1 < CSI ≤ 1.5	Low severity
	1.5 < CSI ≤ 2	Low-moderate severity
	2 < CSI ≤ 2.5	Moderate severity
	2.5 < CSI ≤ 3	Moderate-high severity
	3 < CSI ≤ 4	High severity
	4 < CSI ≤ 5	Very high severity
	5 < CSI	Ultra-high severity
MERMQ	≤0.1	9% probability of toxicity (low priority)
	0.1 < MERMQ ≤ 0.5	21% probability of toxicity (low-medium priority)
	0.1 < MERMQ ≤ 1.5	49% probability of toxicity (medium-high priority)
	1.5 < MERMQ	76% probability of toxicity (high priority)

Major risk assessment methods employed to identify potentially contaminated areas are discussed in the following sections.

2.3.1.1 Single Indices

Single pollution index (PI): given by Hakanson (1980) refers to the ratio between the detection value of the toxic element of interest and the geochemical background value of that element and calculated as follows:

$$\text{PI} = \frac{Ci}{Cb}$$

where *Ci* is the measured concentration of heavy metal in soil and *Cb* is its background concentration. It can be applied to examine which heavy metal is of the maximum threat to soil environment. The lower the index value, the lower the degree of pollution. Its application lies in the calculations of some complex indices e.g., sum of contamination (PI_{sum}), Nemerow pollution index ($PI_{Nemerow}$) (Guan et al. 2014), and the pollution load index (PLI) (Varol 2011).

Threshold pollution index (PI_T): is derived using the ratio between metal concentration and tolerance level of metals (set up by national guidelines) as follows (Weissmannová and Pavlovský 2017):

$$PI_T = \frac{C_i}{C_{TL}}$$

where C_i and C_{TL} are the concentration of heavy metal in soil and tolerance level, respectively. $PI_T > 1$ is considered to be hazardous for health.

Contamination factor (Cf): allows the evaluation of soil contamination by considering the ratio between concentration of heavy metal in soil under study and preindustrial reference level values. It is calculated using the following equation (Hakanson 1980):

$$Cf = \frac{Cm}{C_{p-i}}$$

where *Cm* is the concentration of toxic element in the sample of soil under study and C_{p-i} is the preindustrial reference value.

Enrichment factor (EF): is an indicator to analyze the presence and extent of contaminant deposition on soil surface due to anthropogenic activities. It is calculated using the following equation (Sutherland 2000):

$$EF = \frac{\left(\frac{Ci}{RE}\right)\text{soil}}{\left(\frac{Ci}{RE}\right)\text{background}}$$

where *Ci* is the concentration of toxic element in the sample of soil under study. RE is the concentration of reference element. To determine the expected effect of anthropogenesis on the concentrations of heavy metal in the soil, the concentration of metals characterized by low variability of occurrence, i.e. nondegradable, immobile, and especially stable in soil is used as a reference. Reference elements are generally Al, Ca, Fe, Mn, Sc, or Ti. This calculation normalizes the concentration of metal in topsoil with respect to the concentration of reference element.

Single index of ecological risk factor (E_r): given by Hakanson (1980) can be calculated using the following equation:

$$E_r^i = C_f^i \times T_r^i$$

where C_f^i is the contamination factor of soil for individual element under study and T_r^i is the toxic-response factor of individual toxic element, e.g. 1 for Zn; 2 for Cr; 5 for Cu, Pb, and Ni; 10 for As; 30 for Cd; and 40 for Hg (Hakanson 1980; Xu et al. 2008).

Biogeochemical index (BGI): is designed to evaluate the degree of contamination in O horizon of soil using heavy metal concentration in O and A horizons (Mazurek et al. 2017). There was no such tool for forest and grassland vegetation, and BGI bridges this gap. BGI can be used to estimate the potential of O horizon to sorb heavy metals and reduce phytotoxicity. BGI is calculated as follows:

$$\text{BGI} = \frac{C_n\text{O}}{C_n\text{A}}$$

where C_nO and C_nA are heavy metal content in O and A horizons, respectively. Higher BGI values (i.e. >1) indicate greater ability of O horizon to sorb pollutants. However, BGI only gives an estimate since it does not take the density of soil particles into account.

Geoaccumulation index (I_{geo}): was originally stated by Muller (1969) to determine and compare the present concentration of metals in soil with preindustrial ones. It can be calculated using the following equation:

$$I_{\text{geo}} = \log_2\left[\frac{C_i}{1.5B_i}\right]$$

where C_i is the heavy metal concentration in the site under study and B_i is the reference sample background value. 1.5 is constant, applied to minimize the variability in the background values as a result of lithologic variations in the sediments or lithogenic effect. I_{geo} values can fall in the range of $5 < I_{geo} \le 0$, where moving from lower to higher values shows the increasing level of contamination.

2.3.1.2 Complex/Integrated Indices

Sum of contamination (PI_{sum}): is stated as the sum of all calculated single pollution indices (Gong et al. 2008). It is calculated as follows:

$$\text{PI}_{\text{sum}} = \sum_{i-1}^{n} \text{PI}$$

where PI is the single pollution index and n is the number of heavy metals analyzed. It assesses the overall contamination of group of heavy metals.

Average single pollution index (PI_{avg}): is used to estimate soil quality (Gong et al. 2008; Inengite et al. 2015). It is calculated as follows:

$$PI_{avg} = \frac{1}{n}\sum_{i=1}^{n} PI$$

where n is the number of heavy metals under study. PI_{avg} values >1 depict highly contaminated poor quality soil.

Vector modulus of pollution index (PI_{Vector}): gives an overall assessment of heavy metal contamination of soil (Gong et al. 2008). It is given by the following equation:

$$PI_{vector} = \sqrt{\frac{1}{n}\sum_{i=1}^{n} PI_i^2}$$

where PI_i is the single pollution index of respective metals under study and n is the number of heavy metal species.

Weighted average of pollution index (PIw_{Avg}): defined by Gong et al. (2008), it is a simple statistical tool to study toxicity in a given area. It is calculated as follows:

$$PIw_{Avg} = \sum_{i=1}^{n} W_i PI_i$$

where PI_i is the single pollution index of respective metals under study, n is the number of heavy metal species, and W_i is the class of metal considering the degree of contamination which lies between 1 and 5 (i.e. 5 for Pb, Cd, and Cr; 4 for Ni; 3 for Cu and Fe; 2 for Co and Mn; and 1 for Zn). Higher the value of this index, higher is the toxicity.

Similar to this is **new pollution index (PIN)**, a background enrichment index (Caeiro et al. 2005) which is adapted from PI and also calculated using W_i as follows:

$$PIN = \sum_{i=1}^{n} W_i^2 PI_i$$

Integrated pollution index (IPI): is defined as mean PI value of heavy metals under study and calculated as follows:

$$IPI = \text{mean}\left(PI_i\right)$$

Contamination degree (Cdeg): was stated by Hakanson (1980) to measure the inclusive degree of contamination in soil. It is calculated as follows:

$$\text{Cdeg} = \sum_{i=1}^{n} \text{Cf}$$

where Cf is the contamination factor of heavy metals and n is the number of heavy metals under study.

Modified contamination degree (mCd): is a generalized form of Cdeg (Abrahim and Parker 2008). mCd can be defined as the ratio between sum of all the contamination factor of analyzed pollutants and their number.

$$\text{mCd} = \sum \text{Cf}/n$$

where Cf is the contamination factor and n is the number of elements analyzed.

Pollution load index (PLI): was stated by Tomlinson et al. (1980) and calculated as follows:

$$\text{PLI} = \sqrt[n]{\text{Cf}_1 \times \text{Cf}_2 \times \text{Cf}_3 \times \cdots \times \text{Cf}_n}$$

where Cf is the contamination factor of heavy metal under study.

Potential ecological risk index (PERI): is a comprehensive method and takes metal concentration, toxicity, coactions, and ecological sensitivity into account. It makes utilization of three basic modules, i.e. contamination degree (Cd), toxic-response factor (Tf), and ecological risk index (Er). It is calculated using the following equation (Hakanson, 1980):

$$\text{PERI} = \sum_{i=1}^{n} E_r^i$$

Nemerow's synthetical contamination index (P_N): is a comprehensive index derived from single pollution index (Gong et al. 2008) and is calculated as follows:

$$P_N = \sqrt{\left[\left(1/n\sum P\right)^2 + \left(P_{\max}\right)^2\right]/2}$$

where n is the no of toxic elements analyzed.

Contamination severity index (CSI): stated by Pejman et al. (2015), it helps to determine the threshold of toxicity surpassing which the soil health is affected adversely. CSI employs effects range low (ERL) and effects range median (ERM) values given by Long et al. (1995) (Table 2.3) and is calculated as follows:

$$\text{CSI} = \sum_{i=1}^{n} w_i \left(\left(\frac{C_i}{\text{ERL}_i} \right)^{1/2} + \left(\frac{C_i}{\text{ERM}_i} \right)^2 \right)$$

Table 2.3 ERM and ERL values for elements.

Element	ERL	ERM
Ag	1	3.7
As	8.2	70
Cd	1.2	9.6
Cr	81	370
Cu	34	270
Hg	0.15	0.71
Ni	20.9	51.6
Pb	46.7	218
Zn	150	410

Source: Data from Long et al. (1995).

where w_i is the computed weight of heavy metal under study (i.e. 0.25 for Cd, 0.134 for Cr, 0.075 for Cu, 0.215 for Ni, 0.251 for Pb, and 0.075 for Zn) (Pejman et al. 2015) and C_i is the concentration of each heavy metal under study.

On the basis of the concentrations of heavy metals in consonant with the 10th and 50th percentiles of unfavorable biological effects, three categories are identified. If the concentration is <ERL, adverse effects occur rarely; if the concentration ≥ERL and <ERM, adverse effects occur occasionally; and if the concentration is ≥ ERM, adverse effects occur frequently (Weissmannová and Pavlovský 2017).

Multielement contamination (MEC): stated by Adamu and Nganje (2010), it is used to determine contamination using the concentration of heavy metal in soil and their thresholds stated by Kloke (1979) (i.e. 2 for Hg; 3 for Cd; 20 for As; 100 for Cu, Ni, and Pb; and 300 for Zn). MEC is calculated using the following formula:

$$\text{MEC} = \frac{\left(\frac{C_1}{T_1} + \frac{C_2}{T_2} + \cdots + \frac{C_n}{T_n}\right)}{n}$$

where C is the concentration of each heavy metal under study, T is their threshold value given by Kloke (1979), and n is the number of heavy metals under study.

Mean ERM quotient (MERMQ): is an instrument to assess the harmful effects of heavy metals on soil quality. It was first applied in determining soil quality in reference to corroboration of negative effects (Christophoridis et al. 2009). It is calculated as follows:

$$\text{MERMQ} = \frac{\sum_{i=1}^{n} \frac{C_i}{\text{ERM}_i}}{n}$$

where C_i is the concentration of each heavy metal under study, ERM_i is the ERM value of respective heavy metals as suggested by Long et al. (1995), and n is the number of heavy metals analyzed. MERMQ is applied for identification and prioritizing areas on the basis of hazards due to their soil quality.

Exposure factor (ExF): is a tool to study contamination in an area containing high heavy metal load (Bąbelewska 2010). It is calculated as follows:

$$\text{ExF} = \sum \frac{C_n - C_{\text{av}}}{C_{\text{av}}}$$

where C_n is the concentration of each heavy metal under study and C_{av} is their average content in soil profile.

2.3.2 Human Health Risk Assessment

It is carried out to approximate the possibility and severity of unfavorable health impact on humans due to exposure to contaminants at present and in future. A four-step regime takes into account in order to assess risk to human health

i) Hazard identification: to determine whether the contaminant can lead to an increase in incidence of adverse health effects in humans upon exposure.
ii) Hazard evaluation or dose-response assessment: to determine the probability and extent of unfavorable effects on human health upon exposure to particular amount of contaminant.
iii) Exposure assessment: to determine the extent, frequency, and duration of contaminant exposure to human population. It also considers the exposure pathway and route, i.e. ingestion, inhalation, or dermal contact, followed by contaminant to enter into human body.
iv) Risk characterization: to analyze and integrate the presence or absence of risk based on the above individual components.

Several equations have been designed by regulatory agencies like United States Environmental Protection Agency (US EPA) and are being used by researchers and environmentalists to highlight the risks of contaminant exposure to human health.

2.3.2.1 Estimated Daily Intake (EDI) of Heavy Metal

The estimated daily intake (EDI) values of metal of interest are estimated on the basis of the average content in food items and its daily intake quantity as follows (Antoine et al. 2017):

$$\text{EDI} = \frac{C \times F_{\text{IR}}}{\text{BW}_{\text{a}}}$$

where C is the metal content in the food (mg kg^{-1}), F_{IR} is the daily food ingestion rate (g day^{-1}), and BW_{a} is the reference body weight (kg).

2.3.2.2 Chronic Daily Intake (CDI)

The CDI values via ingestion, inhalation, and dermal absorption routes of contaminated sediments, respectively, can be calculated using the following equations (Kusin et al. 2018):

$$CDI_{Ingest} = \frac{C_{sed} \times IngR \times EF \times ED}{BW_a \times AT} \times CF$$

$$CDI_{Inhale} = \frac{C_{sed} \times InhR \times EF \times ED}{PEF \times BW_a \times AT}$$

$$CDI_{Dermal} = \frac{C_{sed} \times SA \times AF_{sed} \times ABS \times EF \times ED}{BW_a \times AT} \times CF$$

where C_{sed} is the concentration of heavy metal in sediment ($mg\,kg^{-1}$), IngR is the rate of ingestion of sediment ($mg\,day^{-1}$), InhR is the rate of inhalation of sediment ($mg\,cm^{-2}$), PEF is the particle emission factor ($m^3\,kg^{-1}$), SA is the surface area of skin in contact with sediment (cm^2), AF_{sed} is the skin adherence factor of sediment ($mg\,cm^2$), ABS is the dermal absorption factor, EF is the exposure frequency ($days\,year^{-1}$), ED is the exposure duration (years), BW_a is the average body weight (kg), AT is the average time of exposure (days), and CF is the unit conversion factor.

2.3.2.3 Hazard Quotient

CDI_{Ingest} values can be further applied to calculate hazard quotient (HQ) as follows:

$$HQ_{Ing} = \frac{CDI_{Ingest}}{R_f D}$$

where R_fD is the oral reference dose of analyzed heavy metal. R_fD is the maximum acceptable oral dose for toxic substances stated by US EPA. R_fD for As, Al, Cd, Cr, Co, Cu, Fe, Pb, and Zn are 0.0003, 0.0004, 0.001, 0.003, 0.02, 0.0371, 0.7, 0.0035, and 0.3, respectively (US EPA 2012). $HQ \leq 1$ indicates no adverse effects are likely to occur, whereas $HQ > 1$ indicates toxicity to the exposed population.

While R_fD corresponds to oral route, R_fC (reference concentration) is set for inhalation exposure, whereas no reference dose/concentrations are stated for dermal exposure. A separate criterion governs dermal route by identifying chemicals whose dermal exposure exceeds 10% of the oral route and toxicity is calculated using K_p (permeability quotient) (Baynes 2012).

2.3.2.4 Hazard Index

$$HI = \sum HQ = HQ_{Ing} + HQ_{Inh} + HQ_{Der}$$

Alternatively, HI for multiple heavy metals can be calculated as:

$$\mathrm{HI} = \sum_{i=1}^{n} HQ = \mathrm{HQ}_1 + \mathrm{HQ}_2 + \cdots + \mathrm{HQ}_n$$

2.3.2.5 Target Hazard Quotient (Noncarcinogens)

Target hazard quotient (THQ) determines the noncarcinogenic health risks upon exposure to heavy metal of interest. It is assessed as the ratio between exposure of heavy metal and its reference dose (R_fD) and is calculated as per US EPA methodology based on Region III risk-based concentration table as follows:

$$\mathrm{THQ} = \frac{\left(E_{\mathrm{FR}} \times E_{\mathrm{D}} \times F_{\mathrm{IR}} \times C\right)}{R_f D \times \mathrm{BW}_{\mathrm{a}} \times \mathrm{AT}_n} \times 10^{-3}$$

where E_{FR} is the exposure frequency (days year^{-1}) to the heavy metal under analysis, E_{D} is the exposure duration (in years), F_{IR} is the food ingestion rate (g day^{-1}), C is the concentration of heavy metal (mg kg^{-1}), R_fD is the oral reference dose of heavy metal (μg g^{-1} day^{-1}), BW_a is the reference body weight (kg), AT_n is the average time of exposure (365*E_{D}), and 10^{-3} is the unit conversion factor. If THQ value is <1, the exposed population is unlikely to face any health hazards. Whereas THQ > 1 indicates that exposed population faces a potential health risk. This calls for immediate actions and protective measures need to be taken.

Since exposure to more than one pollutants can show additive/interactive effects (Hallenbeck 1993), the cumulative health risks can be determined by summing up THQ values of individual heavy metals and denoted as total THQ (TTHQ) (Ullah et al. 2017) as follows:

$$T\mathrm{THQ} = \mathrm{THQ}_1 + \mathrm{THQ}_2 + \mathrm{THQ}_3 + \cdots + \mathrm{THQ}_n$$

2.3.2.6 Total Lifetime Cancer Risk (TLCR)

Total lifetime cancer risk (TLCR) is the incremental probability of a person to develop any kind of cancer due to exposure to carcinogenic agents over a lifetime. It can be calculated as follows:

$$\mathrm{TLCR} = \mathrm{CDI} \times \mathrm{CSF}$$

where CDI is the chronic daily intake of carcinogenic toxicant and CSF is the cancer slope factor. CSF values for As and Pb are 1.5 and 0.0085 mg kg^{-1} day^{-1}, respectively (US EPA 2012)

$$\sum \mathrm{TLCR} = \mathrm{TLCR}_{\mathrm{Ing}} + \mathrm{TLCR}_{\mathrm{Inh}} + \mathrm{TLCR}_{\mathrm{Der}}$$

where $\mathrm{TLCR}_{\mathrm{Ing}}$, $\mathrm{TLCR}_{\mathrm{Inh}}$, and $\mathrm{TLCR}_{\mathrm{Der}}$ are the total lifetime cancer risk via ingestion, inhalation, and dermal routes, respectively. The permissible threshold value for TLCR is 10^{-6} and 10^{-4} for single and multi-carcinogens, respectively.

2.4 Phytoremediation of Mine Land

Over the years, mining activities have caused destruction of soil texture, disturbance in nutrient cycles, and alteration in microbial communities, ultimately leading to soil damage (Buta et al. 2019).Since prehistoric era, reckless exploitation of mine lands is a leading cause of ecological disturbances worldwide. Countries like China, the United States, Russia, Australia, and India are the leading miners in the world as per International Organizing Committee for World Mining Congress. In order to meet the ever growing demand for mineral resources due to extensive industrialization, urbanization, and population growth, mining activities are being run at different paces in different countries. The end result of mining activities is the same, i.e. degradation of land and loss of soil fertility. Remediation and restoration of degraded mine lands are major goals of sustainable development, concerning over soil structure, fertility, microbial composition, and nutrient cycling so that reclaimed mine land can act as self-sustaining ecosystem (Buta et al. 2019). Reclamation of mine land is accomplished by landscaping, soil amelioration, and revegetation. Recently, Supreme Court of India directed the Center to revisit a decade old "National Mineral Policy, 2008" and prepare a fresh, meaningful, and implementable policy. As a result of which, in their latest order, Ministry of Environment, Forest and Climate Change (MoEFCC) has made re-grassing of mined-out areas mandatory. It is advantageous to commence reclamation immediately after the termination of first phase of mining. Reclamation involves:

i) Landscaping or site preparation involves improving the physical structure of mine land by removal of soil and overburden along with disposal of wastes. Pits and dumps are contoured, and ploughing and mulching are performed to prepare seed bed; an effective drainage and erosion control system is established.
ii) Soil amelioration involves activities tending to overcome conditions such as salinity, toxicity, and deficiency of organic and inorganic nutrients. It also covers treatment of highly acidic or alkaline soils by means of physical and chemical methods to make it suitable for plant growth. For acidic soils, physical treatment includes addition of organic matter as topsoiling helps bury acidic soil deeper, whereas chemical treatments involve addition of lime. In case of alkaline soils, physical treatment involves irrigation to leach excess salts, whereas chemical treatment involves addition of soluble calcium salts (e.g., $CaCl_2$ and $CaSO_4$) and acids or acid formers (S, H_2SO_4, and $Al_2(SO_4)_2$). However, it is more feasible to opt for halophytic plants instead of treating alkaline soils physically and chemically. Mulching is also an effective strategy for soil amelioration.
iii) Revegetation, i.e. establishing vegetation cover over mined area for a range of benefits.

Vegetation reverses degradation by development of extensive root system that acts as soil stabilizer. Plant establishment leads to increase in soil organic matter content and decrease in bulk density and neutralizes soil pH. The root system accumulates nutrients from lower layers and redeposits them into surface layers, rendering them available for microbial breakdown (Mendez and Maier 2008a). In order to design an action plan for reclamation of mine lands via revegetation, investigation of current soil properties is required. Soil quality and its physicochemical properties such as clay fraction, texture, and pH determine its heavy metal accumulation potential (Mohammadi et al. 2022). Additionally, soil quality is one of the key determinants of plant cover that can be supported, and hence influences the success of any phytoremediation project.

Since revegetation is the most accepted and feasible way to control soil erosion and restore soil health, a range of physical, chemical, and biological tests can be carried out in order to choose the kind of vegetation it will support. Soil pH is the most reliable measure of soil acidity indicating mine soil quality and affects soil processes. Factors such as climate and mineral content besides vegetation affect soil pH greatly. While neutral pH (~7) is best suited to most forms of vegetation, pH of mine soil ranges from 6.0 to 7.5 which is preferable for forages (Sheoran et al. 2010). Enhancing soil fertility is an indispensable component of reclamation protocol so as to render soil fit for agricultural use. Organic matter is the principle source of major macronutrients, i.e. nitrogen, phosphorus, and potassium, which are usually found to be deficient in mine soils. Since soils serve as carbon sinks, the content of organic carbon indicates soil quality and depicts land use. Soil in mine lands is generally poor in carbon content mainly because of decomposition of organic carbon due to degradation. Vegetation cover can significantly stimulate carbon sequestration thereby increasing soil organic matter content by adding plant remains to the soil and fixing nitrogen in soil (by using legume crops). Different crops have different contribution to soil reclamation. Shallow rooted grasses are able to improve soil aggregation, whereas deep rooted grasses ease soil compaction. Vetiver grass is a promising candidate for remediation and restoration of mine soil as it has been shown to possess good phytostabilization potential. It can grow on mine soil with its dense mesh like roots and also withstand the harshness of mine soil by enhancing its antioxidant response (Banerjee et al. 2016). Pasture cover significantly enhances soil organic and microbial content besides inorganic nutrients followed by natural grasslands (Buta et al. 2019). Certain wild plants, namely *Amaranthus* spp., *Ageratum conyzoides, Chromolaena odorata, Eleusine inica, Momordica charantia*, and *Sesbania grandiflora*, have found to be potential agents for phytoremediation of mine soil (Hamzah and Priyadarshini 2014). Hence, mixed plantations comprising grasses, shrubs, and trees or practicing crop rotation with leguminous plants and native species are beneficial to a wider range of soil functions due to their adaptation to nutrient deficiency and fast growing traits aiding in restoring fertility and accelerating ecological succession. Phosphorous, though present in both organic and inorganic forms in abundance, is deficient in degraded soils, making it the second major plant growth limiting nutrient. Mine soils are also poor in potassium, the third most important macronutrient due to their structure. The soil fixation ability or availability of these major nutrients in mine lands increases slowly with time; however, repeated fertilizer application is indispensable for faster soil reclamation as it has shown to be beneficial in enhancing vegetation cover by providing sufficient amounts of nutrients required for plant growth.

Besides other indicators, soil enzymatic activity is very sensitive to changes in soil management and indicates soil fertility. Soil enzymes respond to changes in soil management practices even before other indicators. Enzyme activities reflect the microbial oxidative metabolism which is a key reliable indicator of soil health. Enzymes such as saccharase and dehydrogenase are indispensable to nutrient cycles and catalyze key reactions. Revegetation on degraded lands leads to improved enzymatic (catalase, dehydrogenase, and saccharase) activity and soil properties (Buta et al. 2019).

Soil microbes play a vital role in decomposition of soil organic matter. The moisture quantity in soil plays a decisive role in granting dominance to bacteria or fungi in soil system. Fungi tend to dominate over bacteria as they are capable of tolerating a wider range of soil pH. However, soil microbes inhabiting top layers get relocated during mining activities. When the topsoil is removed and stockpiled, soil bacteria can move through the pores with the help of air and water and arrive to top layers. But these activities are detrimental to fungi as their established hypha network disrupts when soil layer is removed. Mycorrhizal associations are known to aid plant establishment, nutrition, and survival under conditions of both abiotic and biotic stress (Diagne et al. 2020). Mycorrhization induces the formation of glomalin-related soil protein and water-stable aggregates and plays a pivot role in improving soil structure (Zhang et al. 2017). Reclamation activities reestablish mycorrhizal densities by adding organic carbon source or plant cover that stimulates microbial activities (Buta et al. 2019).

The release of CO_2 from soil as a result of decomposition of soil organic matter makes soil respiration also a key indicator of soil health. Even though soil respiration indicates soil microbial activity, it is also affected by factors such as temperature, moisture, and pH (Sandor 2010). Since enzymes and microbial metabolism are temperature dependent, soil CO_2 efflux is varied upon temperature fluctuations. Moreover, both drought and high humidity affect microbes, reducing soil respiration in return. Soil pH considerably affects proliferation of microbes, hence affecting soil respiration substantially (Rousk et al. 2009). Forest vegetation followed by pasture and grassland vegetation led to an increase in soil respiration in a mine land reclamation study (Buta et al. 2019). All these indicators are also determinants of effectiveness of reclamation protocol.

In order to select plants to reclaim and remediate degraded lands, drought-resistant, fast growing plants capable in growing nutrient deficiency are selected. In acidic soils, native plants capable of tolerating metals and nutrient deficiency along with possessing high growth and biomass are desirous (Mendez and Maier 2008b).

Indigenous species are preferred over exotics as they are adapted climatically and correspond to functional ecosystem. Moreover, exotics may compete with local flora and act as invasive pests in long run (Sheoran et al., 2010). In a study conducted on phytoremediation potential of native plants at mine tailings site in Ghana, certain native plants *Anthocleista nobilis*, *Alstonia boonei*, *Aspilia africana*, *Costus afer*, and *Marantochloa purpurea* emerged as promising candidates for phytoremediation of heavy metals (Petelka et al. 2019). Similarly, four indigenous plant species *Cynodon dactylon*, *Chromolaena ordorata*, *Chrysopogon zizanioides*, and *Paspalum viginatum* were found to be phytoextractors for cadmium in a mine site (Opoku et al. 2020).

In early phase of revegetation, grasses act as nurse crops as the initial cover must be a diverse, self-sustaining community. Grasses, especially C4 ones are tolerant to drought and other stresses. The fibrous root system of grasses stabilizes soil and enriches it with organic content. Grasses are also capable of competing with weeds. In progressive phase of reclamation, trees act as potential candidates by increasing soil organic content, uptake of nutrients far from reach of herbaceous vegetation, reducing erosion and leaching, decreasing soil acidity, and improving physical properties and biological activities of soil (Sheoran et al. 2010).

Certain species of plants, referred to as metallophytes, are known to inhabit sites such as mine tailings, which are highly polluted with metals and metalloids. Metallophytes are botanical rarities which commonly exist as specialized flora on spoil heaps and mine tailings. Since these are endemic to mine lands, they have well-adapted mechanisms of resistance and tolerance to such high exposure to contaminants. They range between excluders and hyperaccumulators. Metallophytes, both excluders and hyperaccumulators, are interesting systems for evolutionary, physiological, and ecological research of plants. While excluders can exclude metals until a certain toxicity threshold is reached, hyperaccumulators are plants capable of absorbing and accumulating very high concentrations of metals in their tissues with the help of differential gene expression compared to non-hyperaccumulator plants. Majority of the hyperaccumulators are obligate metallophytes, i.e. endemic to metalliferous soils, while others are pseudo metallophytes or facultative metallophytes which hyperaccumulate heavy metals when present on metalliferous soils, but also exist commonly on non-metalliferous soil (Pollard et al. 2014). The extensive distribution and survival efficiency of pseudo metallophytes make them excellent candidates for phytoremediation over rare hyperaccumulators with shallow roots and reduced biomass. Since indigenous plants adapt themselves to metal concentrations over time, they rarely face toxicity issues. Encouraging plant cover by large-scale planting and using seeds of native plants with higher metal accumulation potential is an excellent proposition to remediate mine lands (Favas et al. 2014).

2.5 Conclusion

Soil pollution is an inevitable consequence of mining. An increase in awareness regarding the environment has led to synchronous post-mining reclamation of mine lands as a requisite to mining spectrum. Reclamation is not just limited to above-ground revegetation, but it also takes below-ground biodiversity and soil health into account.

In order to establish correct theoretical basis to interpret soil condition accurately, several indices have been proposed at individual and integrated level. The employment of these indices for both human health and ecological risk assessment leads to a better understanding of soil health and highlights the hazards associated with human health upon consumption.

To restore the ecological integrity of mine land, revegetation is the most feasible, environmentally pleasing strategy to protect land against erosion and degradation. To implement an effective reclamation strategy, selection of suitable plants plays the most crucial role. Plant selection should be based on the ability to establish and survive in the given soil type. In this context, indigenous (native) species emerge as the most suitable candidates for reclamation. Further, practices such as mixed cropping (including leguminous crops) lead to soil enrichment due to associations with nitrogen-fixing bacteria and fungi, which ultimately aid in soil fertility.

References

Abrahim, G.M.S. and Parker, R.J. (2008). Assessment of heavy metal enrichment factors and the degree of contamination in marine sediments from Tamaki Estuary, Auckland, New Zealand. *Environmental Monitoring Assessment* 136: 227–238.

Adamu, C.L. and Nganje, T.N. (2010). Heavy metal contamination of surface soil in relationship to land use patterns: a case study of Benue State, Nigeria. *Materials Sciences and Applications* 1: 127–134.

Agyeman, P.C., John, K., Kebonye, N.M. et al. (2021). Human health risk exposure and ecological risk assessment of potentially toxic element pollution in agricultural soils in the district of Frydek Mistek, Czech Republic: a sample location approach. *Environmental Sciences Europe* 33: 137.

Antoine, J.M.R., Fung, L.A.H., and Grant, C.N. (2017). Assessment of the potential health risks associated with the aluminium, arsenic, cadmium and lead content in selected fruits and vegetables grown in Jamaica. *Toxicology Reports* 4: 181–187.

Bąbelewska, A. (2010). The impact of industrial emissions on heavy metal and sulphur contamination level within the area of the projected Jurassic National Park. *Pra̦dnik Studies and Reports of the Prof Władysław Szafer Museum* 20: 135–145.

Banerjee, R., Goswami, P., Pathak, K., and Mukherjee, A. (2016). Vetiver grass: an environment clean-up tool for heavy metal contaminated iron ore mine-soil. *Ecological Engineering* 90: 25–34.

Baragaño, D., Forján, R., Aguado, J.M.M. et al. (2019). Reuse of Dunite mining waste and subproducts for the stabilization of metal(oid)s in polluted soils. *Minerals* 9: 481.

Baynes, R.E. (2012). Quantitative risk assessment methods for cancer and noncancer effects. In: *Toxicology and Human Environments, Progress in Molecular Biology and Translational Science*, vol. 112 (ed. E. Hodgson), 259–283. Elsevier.

Buta, M., Blaga, G., Paulette, L. et al. (2019). Soil reclamation of abandoned mine lands by revegetation in Northwestern part of Transylvania: a 40-year retrospective study. *Sustainability* 11: 3393. https://doi.org/10.3390/su11123393.

Caeiro, S., Costa, M.H., Ramos, T.B. et al. (2005). Assessing heavy metal contamination in Sado estuary sediment: an index analysis approach. *Ecological Indicators* 5 (2): 151–169. https://doi.org/10.1016/j.ecolind.2005.02.001.

Chen, H.Y., Teng, Y.G., Lu, S.J. et al. (2015). Contamination features and health risk of soil heavy metals in China. *Science of the Total Environment* 512: 143–153.

Christophoridis, C., Dedepsidis, D., and Fytianos, K. (2009). Occurrence and distribution of selected heavy metals in the surface sediments of Thermaikos Gulf, N. Greece. Assessment using pollution indicators. *Journal of Hazardous Materials* 168 (2–3): 1082–1091. https://doi.org/10.1016/j.jhazmat.2009.02.154.

Diagne, N., Ngom, M., Djighaly, P.I. et al. (2020). Roles of arbuscular mycorrhizal fungi on plant growth and performance: importance in biotic and abiotic stressed regulation. *Diversity* 12 (10): 370. https://doi.org/10.3390/d12100370.

Ensley, B.D. (2000). Rationale for use of phytoremediation. In: *Phytoremediationof toxic metals. Using plants to clean up the environment* (ed. I. Raskin and B.D. Ensley), 3–11. New York: John Wiley & Sons, Inc.

Fagbot, E.O. and Olanipekun, E.O. (2010). Evaluation of the status of heavy metal pollution of soil and plant (*Chromolaena odorata*) of Agbabu Bitumen Deposit Area, Nigeria. *American-Eurasian Journal of Scientific Research* 4: 241–248.

Favas, P.J.C., Pratas, J., Varun, M., D'Souza, R., and Paul, M.S. (2014). Phytoremediation of soils contaminated with metals and metalloids at mining areas: potential of native flora. http://dx.doi.org/10.5772/57469

Fazekaš, J. (2018). Manazment of Environmentally Burdened Areas. PhD Thesie, University of Prešov in Prešov, Prešov, Slovakia.

Fazekašová, D. and Fazekaš, J. (2020). Soil quality and heavy metal pollution assessment of iron ore mines in Nizna Slana (Slovakia). *Sustainability* 12: 2549. https://doi.org/10.3390/su12062549.

Geelani, S.M., Bhat, S.J.A., Geelani, S.H. et al. (2013). Mining and its impacts on environment with special reference to India. *International Journal of Current Research.* 5 (12): 3586–3590.

Gong, Q., Deng, J., Xiang, Y. et al. (2008). Calculating pollution indices by heavy metals in ecological geochemistry assessment and a case study in parks of Beijing. *Journal of China University of Geosciences* 19: 230–241.

Guan, Y., Shao, C., and Ju, M. (2014). Heavy metal contamination assessment and partition for industrial and mining gathering areas. *International Journal of Environmental Research and Public Health* 11: 7286–7303.

Hakanson, L. (1980). An ecological risk index for aquatic pollution control: a sedimentological approach. *Water Research* 14: 975–1001.

Hallenbeck, W.H. (1993). *Quantitative Risk Assessment for Environmental and Occupational Health*. Chelsea: Lewis.

Hamzah, A. and Priyadarshini, R. (2014). Identification of Wild Grass as Remediator Plant on Artisanal Gold Mine Tailing. *Plant Science International* 1: 33–40.

Hu, Y., Liu, X., Bai, J. et al. (2013). Assessing heavy metal pollution in the surface soils of a region that had undergone three decades of intense industrialization and urbanization. *Environmental Science and Pollution Research* 20: 6150–6159.

Inengite, A.K., Abasi, C.Y., and Walter, C. (2015). Application of pollution indices for the assessment of heavy metal pollution in flood impacted soil. *International Research Journal of Pure & Applied Chemistry* 8: 175–189.

Jomova, K., Jenisova, Z., Feszterova, M. et al. (2011). Arsenic: toxicity, oxidative stress and human disease. *Journal of Applied Toxicology* 31: 95–107.

Kelepertzis, E. (2014). Accumulation of heavy metals in agricultural soils of Mediterranean: Insights from Argolida basin, Peloponnese, Greece. *Geoderma* 221–222: 82–90.

Kierczak, J., Pedziwiatr, A., Waroszewski, J., and Modelska, M. (2016). Mobility of Ni, Cr and Co in serpentine soils derived on various ultrabasic bedrocks under temperate climate. *Geoderma* 268: 78–91.

Kloke, A. (1979). *Content of Arsenic, Cadmium, Chromium, Fluorine, Lead, Mercury, and Nickel in Plants Grown on Contaminated Soils*, 51–53. Geneva: United Nations-ECE Symposium.

Kowalska, J., Mazurek, R., Ga̧siorek, M. et al. (2016). Soil pollution indices conditioned by medieval metallurgical activity: a case study from Krakow (Poland). *Environmental Pollution* 218: 1023–1036.
Kowalska, J.B., Mazurek, R., Gąsiorek, M., and Zaleski, T. (2018). Pollution indices as useful tools for the comprehensive evaluation of the degree of soil contamination–a review. *Environmental Geochemistry and Health* 40: 2395–2420.
Kusin, F.M., Azani, N.N.M., Hasan, S.N.M.S., and Sulong, N.A. (2018). Distribution of heavy metals and metalloid in surface sediments of heavily mined area for bauxite ore in Pengerang, Malaysia and associated risk assessment. *Catena* 165: 454–464.
Long, E.R., MacDonald, D.D., Smith, L., and Calder, F.D. (1995). Incidence of adverse biological effects within ranges of chemical concentrations in marine and estuarine sediments. *Environmental Management* 19: 81–97.
Ma, T., Zhou, W., Yang, X. et al. (2019). Risk assessment of contamination by potentially toxic metals: a case study in the vicinity of an abandoned Pyrite Mine. *Minerals* 9: 783. https://doi.org/10.3390/min9120783.
Mazurek, R., Kowalska, J., Gasiorek, M. et al. (2017). Assessment of heavymetals contamination in surface layers of Roztocze National Park forest soils (SE Poland) by indices of pollution. *Chemosphere* 168: 839–850.
Mendez, M.O. and Maier, R.M. (2008a). Phytoremediation of mine tailings in temperate and arid environments. *Reviews Environmental Science and Biotechnology* 7: 47–59.
Mendez, M.O. and Maier, R.M. (2008b). Phytostabilization of mine tailings in arid and semiarid environments – an emerging remediation technology. *Environmental Health Perspectives* 116 (3): 278–283.
Mileusnić, M., Mapani, B.S., Kamona, A.F. et al. (2014). Assessment of agricultural soil contamination by potentially toxic metals dispersed from improperly disposed tailings, Kombat mine, Namibia. *Journal of Geochemical Exploration* 144: 409–420.
Mohammadi, A., Mansour, S.N., Najafi, M.L. et al. (2022). Probabilistic risk assessment of soil contamination related to agricultural and industrial activities. *Environmental Research* 203: 111837.
Muller, G. (1969). Index of geoaccumulation in sediments of the Rhine River. *GeoJournal* 2: 108–118.
Nedyalkova, M. and Simeonov, V. (2019). Chemomertic risk assessment of soil pollution. *Open Chemistry* 17: 711–721.
Obiora, S.C., Chukwu, A., and Davies, T.C. (2016). Heavy metals and health risk assessment of farmland soils and food crops around Pb-Zn mining localities in Enyigba, southeastern Nigeria. *Journal of African Earth Sciences* 116: 182–189.
Opoku, P., Gikunoo, E., Arthur, E.K., and Foli, G. (2020). Removal of selected heavy metals and metalloids from an artisanal gold mining site in Ghana using indigenous plant species. *Environmental Management & Conservation* 6 (1): 1840863. https://doi.org/10.1080/23311843.2020.1840863.

Pashkevich, M.A. (2017). Classification and environmental impact of mine dumps. In: *Assessment, Restoration and Reclamation of Mining Influenced Soils*, 1–32. Academic Press https://doi.org/10.1016/B978-0-12-809588-1.00001-3.

Pejman, A., Gholamrez Nabi, B., Saeedi, M., and Baghvanda, A. (2015). A new index for assessing heavy metals contamination in sediments: a case study. *Ecological Indicators* 58: 365–373.

Petelka, J., Abraham, J., Bockreis, A. et al. (2019). Soil heavy metal(loid) pollution and phytoremediation potential of native plants on a former Gold Mine in Ghana. *Water, Air, & Soil Pollution* 230: 267.

Pollard, A.J., Reeves, R.D., and Baker, A.J. (2014). Facultative hyperaccumulation of heavy metals and metalloids. *Plant Science* 217–218: 8–17. https://doi.org/10.1016/j.plantsci.2013.11.011.

Reimann, C. and Garret, R.G. (2005). Geochemical background: concept and reality. *Science of the Total Environment* 350: 12–27.

Rousk, J., Brookes, P.C., and Baath, E. (2009). Contrasting soil pH effects on fungal and bacterial growth suggest functional redundancy in carbon mineralization. *Applied and Environmental Microbiology* 75: 1589–1596.

Sandor, M. (2010). Soil respiration: concept and measurement methods. *ProEnvironment* 3 (5): 54–57.

Sheoran, V., Sheoran, A.S., and Poonia, P. (2010). Soil reclamation of abandoned mine land by revegetation: a review. *International Journal of Soil, Sediment and Water* 3 (2): Article 13.

Sołek-Podwika, K., Ciarkowska, K., and Kaleta, D. (2016). Assessment of the risk of pollution by sulfur compounds and heavy metals in soils located in the proximity of a disused for 20 years sulfur mine (SE Poland). *Journal of Environmental Management* 180: 450–458.

Sutherland, R.A. (2000). Bed sediment-associated trace metals in an urban stream, Oahu, Hawaii. *Environmental Geology* 39: 611–627.

Tamene, F.D. and Seyoum, L. (2015). Temporal and spatial variations on heavy metals concentration in River Mojo, Oromia State, East Ethiopia. *International Journal of Science and Research (IJSR).* 4 (1): 1424–1432.

Tomlinson, D.L., Wilson, J.G., Harris, C.R., and Jeffrey, D.W. (1980). Problems in assessment of heavy metals in Estuaries and the formation of Pollution Index. *Helgoland Marine Research* 33 (1): 566–575.

Ullah, A.K.M.A., Maksud, M.A., Khan, S.R. et al. (2017). Dietary intake of heavy metals from eight highly consumed species of cultured fish and possible human health risk implications in Bangladesh. *Toxicology Reports* 4: 574–579.

US Environmental Protection Agency (US EPA) (2012). *Integrated Risk Information System of the US Environmental Protection Agency*. US EPA.

Varol, M. (2011). Assessment of heavy metal contamination in sediments of the Tigris River (Turkey) using pollution indices and multivariate statistical techniques. *Journal of Hazardous Materials* 195: 355–364.

Weissmannová, H.D. and Pavlovský, J. (2017). Indices of soil contamination by heavy metals – methodology of calculation for pollution assessment (minireview). *Environmental Monitoring and Assessment* 189: 616.

Xu, Z.Q., Ni, S.J., Tuo, X.G., and Zhang, C.J. (2008). Calculation of heavy metals toxicity coefficient in the evaluation of potential ecological risk index. *Environmental Science Technology* 31 (2): 112–115.

Yang, Q., Li, Z., Lu, X. et al. (2018). A review of soil heavy metal pollution from industrial and agricultural regions in China: pollution and risk assessment. *Science of the Total Environment* 642: 690–700.

Zhang, Y.C., Wang, P., Wu, Q.H. et al. (2017). Arbuscular mycorrhizas improve plant growth and soil structure in trifoliate orange under salt stress. *Archives of Agronomy and Soil Science* 63 (4): 491–500. https://doi.org/10.1080/03650340.2016.1222609.

Zhou, S., Zhou, K., Cui, Y. et al. (2015). Identifying geochemical anomalies according to the content and coefficient variance of trance elements. *Scientia Geologica Sinica* 50: 1014–1022.

3

Bio-Geotechnologies in Mine Land Restoration

Abstract

This chapter emphasizes on post-mining reclamation techniques of disturbed lands as a result of mining. As mining poses serious environmental threats and deleterious effects on the landscape, reclamation is a challenge and requires elaborate planning much before mines are opened. In this chapter, a detail regarding proper reclamation strategies including probable bio-geotechnologies including phytoremediation, nanotechnology, silviculture, transgenics effective in mine land restoration is presented. Land reclamation has been a matter of concern to counter the negative impacts of mining and address the issues of landscape remodeling. The chapter highlights the major techniques beneficial for the strategic and sustainable reclamation of disturbed mine lands and to return the aesthetic pleasure of the land site which can be commercially and agronomically productive and suitable for future use. The basic proposal for land reclamation requirements and various methods implied for mine land restoration are discussed elaborately in this chapter.

Keywords *Coal mine; restoration; microbes; phytoremediation; phytodegradation; phytoextraction; amendments*

CONTENTS

Eco-Restoration of Mine Land, First Edition. Vimal Chandra Pandey, Jitendra Ahirwal, Roopali Roychowdhury, and Ritu Chaturvedi.

3.1 Introduction

Metal extraction and other mining industries have notable contributions toward a country's economy by job generations as well as revenue generation through exports. However, the increasing power demand is resulting in excessive coal digging leading to acid leachate generation by coal wastes, rock wastes, burrowed discard, and coal crumbles which causes "acid mine drainage (AMD)." As a matter of environmental concern, these leachates need to be managed and restored to avert adverse impacts. AMD is principally accountable for environmental loss due to highly toxic metal leachates, percolation of which can cause ground and surface water contamination. Sulfides in the mined rocks when excavated are exposed to water and oxygen which encounter oxidized reactions catalyzed by bacterial actions increasing the Fe, sulfate, and other toxic metal concentrations in groundwaters in mine voids (Eq 3.1)

$$FeS_2(s) + 3.75O_2 + 3.5H_2O \rightarrow Fe(OH)_3(s) + 2SO_4^{2-} + 4H^+ \tag{3.1}$$

Mining venture accelerates environmental degradation through acidic effluence, degradation of ecosystem causing loss of greenery, alteration in the hydrological cycle and soil attributes, hampering ecological balance, and loss of biodiversity. Impacts such as reduced agricultural land, lowland ponding, road cracks, and house cracks are also noticed (Balat 2007). According to a study done (Groninger et al. 2007), mining activities have been found to affect about 1 060 000 hectares of forests and 263 000 hectares of grasslands (Groninger et al. 2007). Additionally, surface mining contributes immensely to the air, water and soil pollution which is unfavorable for human health and causing diseases such as hypertension, pulmonary disorder, lung cancer, and kidney diseases (Palmer et al. 2010). The wastes discarded from mining activities in the form of rocks and tailings heap upon the facet and prohibit land use, covering huge areas causing degradation and unfit for use (Li 2006). Soil annihilation is one of the most urgent ecological effects of opencast mining exercises. Throughout expelling the ideal mineral component, unique substratum becomes bygone or gets covered by squanders. During excavation, the topsoil must be saved to support the basic wellspring of seed and supplements and has to be safeguarded for use in recovery.

Lack of arable land, pollution, soil degradation, and loss of biodiversity are some of the prime aspects which require immediate attention. However, mining processes cannot be fully curtailed keeping in view the progress of a nation. Coal resource exploitation and environmental management awareness are needed to ensure sustainability. Thus, reclamation challenges and efficient land use are the key features to be addressed. Most of the nations are adopting to mine reclamation patterns using "engineering designs" which includes site restoration, amendment of soil quality, restructuring, and revegetation (Gao et al. 1998; Wang et al. 2001; Skleniˇcka et al. 2002). The most commonly accepted surface mining strategy includes open-pit and strip mining, and quarrying. Resultingly, the disturbance of the topsoil influences the humus, biota, plant life, and surface water, thereby, impacting a wide range of land use. Moreover, consistent digging of the subsurface water table will hamper surface groundwater (Chamber of Mines of South Africa 2008).

Surface minings are usually carried out on a large scale and lead to a massive disruption of land, air, and water quality. It also degrades land and simultaneously is responsible for massive pollution at mining sites. Health hazards come hand in hand with mining activities. The area in and around the mining sites needs careful study before mining activities are conducted to plan and cope up with the post-mining reclamation techniques. Removal, storing, and final restoration of the soil need attention which might otherwise pose threats during the reclamation process (Rimmer and Younger 1997). Kavourides et al. (2002) have listed few counter effects of mining on the natural habitat which comprise requirements of large farmlands for digging and disposal, a transformation of land area, perturbing vegetation and wildlife, disrupting ground and surface water balance, resettlements of residential areas, and addition of environmental pollution to name a few.

Many ecological aspects have to be taken into consideration before planning any mine reclamation process. As soil is the major criteria that have to be taken into consideration before any action, certain essential facts such as pH, fertility, porosity, microbiota, and soil nutrients have to be kept in mind (Sheoran et al. 2010). Ecological restoration of mine land areas needs precise planning apart from engineering actions. Concepts of building parks, residential or business areas, or reclamation into agricultural lands can be planned in the restoration schedule for abandoned mines. However, this depends on the location and site conditions of the proposed area.

Reclamation of mining areas or deserted mines is a complex process and preplanning measures need to be adopted for a smooth remedial process. The aptness of any land can be achieved through proper introspection of quality measures such as climate, topography, social and economic conditions, and environment. Because of this, such areas which do not have favorable conditions meeting topography, position, and climate, specifically cannot be restored into areas of economic interests. Such areas can only be reclaimed as woodlands; however, mining sites adjoining urban areas or approaching city by lanes can be developed and restored as a business, commercial, or residential structure. The multidimensional utilization of a specific mining site is, however, quite unusual as the land structure would not support both forest and commercial use simultaneously. Hence preplanning with concerns is essential before revamping (Mendoza 2000).

The practice of ecological restoration of mine areas has to be conceptualized based on the ecosystem services; however, it is seldom used in practice during the planning of mine site restoration (Liu et al. 2008; Larondelle and Haase 2012).

3.2 Potential Approaches for Mine Land Restoration

3.2.1 Phytoremediation

Phytoremediation technique has received special mention in different works of literature. The term "phytoremediation" can be divided as (Greek) "Phyto for plant, and remedium (Latin) to remediate or remove an evil" (Erakhrumen and

Table 3.1 Different definitions of phytoremediation.

S. No.	Scientist	Definition
1.	Negri et al. (1996)	"The engineered use of the green plant to remove, contain, or render harmless such environmental contaminants as heavy metals, trace elements, organic compounds, and radioactive compounds in soil or water. This definition includes all plant-influenced biological, chemical, and physical processes that aid in the uptake, sequestration, degradation, and metabolism of contaminants, either by plants or by the free-living organisms that constitute the plant rhizosphere"
2.	Rodriguez et al. (2005)	"The use of plants to remediate toxic chemicals found in contaminated soil, sludge, sediment, groundwater, surface water, and wastewater"
3.	US EPA (2000)	"Phytoremediation is the name given to a set of technologies that use different plants as a containment, destruction, or an extraction technique. Phytoremediation is an emerging technology that uses various plants to degrade, extract, contain, or immobilize contaminants from soil and water"
4.	Moreno et al. (2008)	"The use of plants to improve degraded environments"
5.	Prasad (2003)	"The use of plants, including trees and grasses, to remove, destroy or sequester hazardous contaminants from media such as air, water, and soil"
6.	Liu et al. (2000)	"An emerging technology using specially selected and engineered metal accumulating plants for environmental cleanup"
7.	Bhattacharya et al. (2006)	"The use of vascular plants to remove pollutants from the environment or to render them harmless"
8.	Van Ginneken et al. (2007)	"Phytoremediation, in general, implies the use of plants (in combination with their associated microorganisms) to remove, degrade, or stabilize contaminants"

Agbontalor 2007). However, many researchers have defined phytoremediation separately (Table 3.1). In general term, phytoremediation engages plant species to eliminate organic and inorganic contaminant from adulterated or degraded soil or water for a fresh environment.

Phytoremediation is a newly emerging technique in the area of science that involves vegetation including grasses, herbs, or woody species to redevelop damaged land and water (ground and wastewater) while removing different pollutants namely metals, trace elements, organic and radioactive components, and other metalloids. The absorption of metal from the soil is initiated by the plant and is mainly the responsibility of the roots to sequester them. Later, it is transported and bioaccumulated in the vegetative parts (shoot and leaves). However, certain metal pollutants experience a change in the state through phytodegradation, phytotransformation, or phytovolatilization thereby reducing the metal load of the soil as well as the plant. Being pocket-friendly and ecologically safe, this technique has reported success by many scientists in the treatment of polluted soil having multiple pollutants. Thus, this method is getting acclaimed in the restoration of FA dump sites and becoming popular as conventional treatment options such as incineration and bioslurry composting are not further required (Mpofu et al. 2013).

Mining sites can be reclaimed post operation using engineering techniques of excavation to mitigate pollution caused by metal contaminants (Tordoff et al. 2000). This sometimes includes physical, biological, and chemical techniques (Wood 1997). However, for large areas phytoremediation proves to be beneficial and a cheap option to immobilize the contaminants in the area (Sheoran et al. 2010). Areas experiencing fewer contaminants or show less influence of pollution can be kept isolated and provided with chemical, thermal, or pyrometallurgical treatments along with in situ soil douche and chemical percolation (Mulligan et al. 2001; Lottermoser 2010). For in situ site reclamation, the soil amendment technique has to be focused upon, which can involve biochemical treatments (Sydnor and Redente 2002; Yang et al. 2006). Hettiarachchi et al. (2001) emphasize the tapping of phosphorus medium to be potent in lowering lead concentrations along with a vegetation cover on the land (Mendez and Maier 2008).

Typically added amendments such as clays and alkalescent substitutes are suitable for zinc; however, for chromium contamination, iron and other organic matters are preferable. For arsenic remediation, manganese and iron oxides are promising along with suitable vegetation as suggested by Kumpiene et al. (2008).

As reports of harmful effects of phosphates and alkali metals are prominent on soils with arsenic, similar effects are reported in the case of manganese oxides and alkalis on chromium-incited soils (Kim and Dixon 2002; Rai et al. 2004). Freitas et al. (2013) have documented an upsurge of upto 14 times in Pb uptake through phytoaccumulation in vetiver grass species and maize if supplemented with chelating agents such as citric acid (approximately 40 mmol kg^{-1}). Mine lands are loaded with pollutants that may be organic and/or inorganic (Karbowska 2016). This is a result of the leaching of harmful metals in and around the soil areas which remain untreated and tend to make the soil nutrient deficient and unfit for immediate use. To facilitate restoration, the mine soils need proper analysis in context to the texture, pH, salinity, metal content, organic matter, and root exudates. Rehabilitation practices have long been applied differently to make abandoned mining sites eco-friendly and reusable. Treatment options for mining discards as well as AMD have been tried which has produced mixed results in terms of usability. Sustainable recovery of these sites has been tried for removing heavy metal contamination (Wuana and Okieimen 2011). Precisely, desulfurization by H_2O_2 along with cement paste application; "zerovalent iron barriers; electrodialytic remediation; options using marble, fly ash (FA), some carbonates, lime treatment, soil cleaning, and precipitation; metal-binding amendments; polymer treatment for metal recovery from mining refuse, etc., along with nano remediation treatment of AMD; metal retrieval from waste by microbes-assisted bioremediation, remediation by biochar application; 'phytostabilization' of mine wastes; cover cropping; and mulching over mine tailings;" are the ones which need special mention. Initiating a vegetation cover on these metal-rich land requires identification of specific "metal excluder" or tolerant plant species which would effectively extract and stabilize metal movement in the soil (Figure 3.1) through the phenomenon of phytoremediation (González et al. 2019).

Figure 3.1 explains the essentials of the phytoremediation technique (USEPA 2000; Prasad 2003) such as phytoextraction and stabilization which

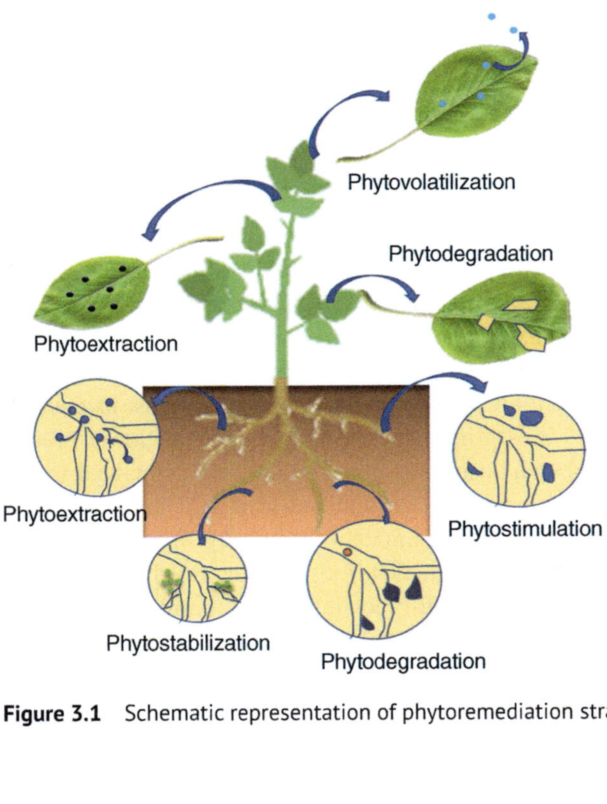

Figure 3.1 Schematic representation of phytoremediation strategies.

involves inorganic pollutants, phytodegradation also termed as phytotransformation, rhizofiltration, and rhizodegradation for carbon-based contaminants. In the phytostabilization process, the plant enzymes from the roots help in immobilizing the pollutants in the soil reserve, thereby prohibiting further transfer. Earlier researchers have documented that phytostabilization is efficient in removing soil and groundwater contaminants either by absorption and precipitation on roots or inside the rhizosphere. This technique holds good for metal and organic pollutants in soil and sludges. Specific plant species called "accumulators/hyperaccumulators," however, emphasize on the assimilation of harmful pollutants in their harvestable parts which is termed as phytoextraction. This method is quite promising for metals, nonmetals as well as radionuclides present in soil or slurry (USEPA 2000; Prasad 2003). Plant species capable of extracting pollutants from affected soil or water media, absorbing and further passing it off as vapor into the atmosphere, exhibit the process of phytovolatilization. On the other hand, the phytodegradation process involves the disintegration of toxins assembled by the plants via different processes and compounds produced by the plant itself (USEPA 2000; Prasad 2003).

Land reclamation through foliage and alterations in the soil quality can be enhanced by aided phytostabilization techniques requiring metal accumulator plant species in the marked area. This plant cover will not only recuperate soil quality but also minimize metal mobility by absorption or adsorption process. Moreover, the soil fertility, microbiota, and texture would also revive in future (Bolan et al. 2011). Metal accumulation is mainly into plant roots or rhizosphere apart from others. Thus, phytomanagement ensures the safe removal of metals by translocation in the metal tolerant plant parts, thereby minimizing the risk of being dispersed or leached by soil, wind, or water (Bolan et al. 2011). Hence, phytostabilization can be counted on as an effective eco-friendly and economic alternative in restoring abandoned mine wasteland in comparison to conventional techniques.

Several studies indicate that mining sites are laden with fly ash wastes and mine tailings when shielded by vegetation, mitigate environmental pollution as it controls erosion by wind, bind with pollutants, and help decreasing toxicity and mobility, thereby minimizing the possibility of metal contamination via the food chain. (DŽeletović and Filipović 1995; Pavlović et al. 2004; Djurdjević et al. 2006; Mitrović et al. 2008; DŽeletović et al. 2009; Haynes 2009; Pandey et al. 2015a, b, 2016a; Gajić et al. 2016). Accordingly, eco-restoration of debased destinations is the way toward recharging biological system steadiness and flexibility after pressure or aggravation as for its well-being, trustworthiness, and manageability (SER 2002), introducing a vital topic in natural science and environmental designing.

In eco-restoration projects, the indigenous plant species growing over some time in similar areas of contamination have attuned to the conditions and therefore can prove to be the best replacements as valuable restoration aids (Richards et al. 1998; Dorner 2002). These selected plant species over a while would build up a community on the abandoned sites and would require the least human interventions for monitoring. To develop a phyto cover on the desolate land, certain self-managing plants are proposed which are "easy for plantation, have a deep root system, fast-growing, nitrogen fixers, stress-tolerant, and are helpful in nutrient pathways" (Maiti 2012).

Human efforts of planting shrubs and trees through seeding the mixture of grasses and legumes are fruitful in establishing a self-sustaining green cover leading to effective restoration (Pavlović et al. 2004; Mitrović et al. 2008; Kostić et al. 2012; Maiti 2012; Gajić et al. 2016). Earlier documented grass and legumes species provide suitable vegetation on degraded lands in a temperate climate. In the tropical regions, a variety of grasses and trees are prominent though. Apart from these, certain leguminous species also show effective results in the revegetation of barren lands. Table 3.2 shows the plant varieties effective in the revegetation of degraded sites.

Legumes are typical nitrogen fixers in the presence of bacteria and can provide the plant with essential nitrogen in specific abandoned lands filled with fly ash and mine spoils. On the other hand, different cereal crops and grasses serve as potential colonizers in the revegetation process, arresting soil erosion and improving soil quality over a while (Maiti 2012; Gajić et al. 2016).

Eventually, during the time frame, the colonizing plants would form thick vegetation cover on the abandoned sites. Annual and biennial plants are the presiding ones over disposed of soils. However, the annuals are momentary having well growth along with the production of large seeds (example *Erigeron, Chenopodium*) during their growth period, and biennials are producers of ample seeds along with a life span of few years. Biannuals can be "competitive ruderals (C-R) and stress-tolerant ruderals (S-R)" as suggested by Grime (1979) (for example *Daucus carota, Tragopogon dubius, Verbascum phlomoides, Oenothera biennis*). Proliferation and growth from the seeds, stolons, rhizomes or bulbs, and tubers are the common asexual reproduction modes in herbaceous perennial plants, which store reserves for unfavorable situations and are termed as "stress-tolerant competitors (C-S), stress-tolerant ruderals (S-R), competitive ruderals (C-R)," and "C-S-R." The rhizomes serve the purpose of up taking reserves for existence during harsh conditions and connect projecting shoots (*Phragmites communis, Calamagrostis epigejos, Festuca rubra, Cirsium arvense, Epilobium*) (Grime 1979).

3.2.2 Phytomining

Plants react differently in the presence of metal stress in the soils. Most of them show sensitivity to metal contamination, whereas others show resistance and tolerate stress by absorbing metal through uptake mechanisms, which are then

Table 3.2 Plant varieties effective in revegetation of degraded sites.

Region	Plant varieties				Source
	Legumes	Grasses	Cereals	Trees	
Temperate	*Medicago*, and *Vicia*	*Festuca, Lolium, Agropyron, Poa*		*Poplar, Robinia, Salix, Alnus, Betula*, and *Acer*	Prasad (2006)
Tropical		*Chloris, Pennisetum, Vetiveria, Paspalum, Cynodon*		*Leucaena, Acacia*, and *Sesbania* (legumes) and *Prosopis, Eucaliptus* and *Ricinus*	Prasad (2006)
		Dactylis glomerata, F. rubra, Poa pratensis, Lolium multiflorum, Lolium perenne, Arrhenatherum elatius, Phleum pretense	*Secale cereale, Avena sativa, Hordeum sativum, Triticum aestivum,*	*Acacia, Acer, Azadirachta, Albizia, Amorpha, Cassia, Dalbergia, Eucalyptus, Fraxinus, Grevillea, Leucaena, Melia azedarach, Morus, Platanus, Pongamia pinnata, Populus, Phyllanthus emblica, Rosa, Rubus, Salix, Tamarix, Tectona grandis*	Cheung et al. (2000), Maiti (2012), Gajić et al. (2018)

assembled in the vegetative parts of the plants (Barcelo et al. 1994). When the metal concentrations accumulated by a plant are defined to be more than the normal limit, the specific plant species is termed to be a hyperaccumulator plant compared to others in the same area. Anderson et al. (1999a) have enlisted some of these plant species as hyperaccumulators. Normally, hyperaccumulator plants are stress-tolerant to grow in disturbed lands. As mentioned earlier, specific plants called hyperaccumulators, which are capable of efficient uptake of specific or multiple metals, have been recognized. These are effective in collecting about 0.5–1% of their dry weight in metals. Major species include *Alyssum bertolonii* (1% or 10 000 mg kg^{-1} Ni) and *Brassica juncea* (Au). Baker et al. (2000) defines "hyperaccumulators as higher plant varieties whose shoot contains >10 000 mg Mn and Zn/kg, >1 000 mg Pb, Ni, and Cu/kg, >100 mg Cd/kg (dry wt.)" (Baker et al. 2000).

Extraction or mining of metals from the metal-induced soils is done through hyperaccumulators that deposit metals in plant parts which is allowed suitable growth and then harvested. This material is then dried and charred to ashes and further treated to extract metals undergoing refining processes such as electrowinning and acid dissolution (Robinson et al. 1999) for further use. Phytomining is a powerful technique of metal remediation from metal-contaminated mine sites and retrieving the amount of metals from the plants (Anderson et al. 2003). Phytomining is coined as "the production of a 'crop' of a metal by growing high-biomass plants that accumulate high metal concentrations" (Brooks et al. 1998). Hyperaccumulators are naturally occurring plants and in others who show similar qualities, the property can be induced. Plant species fulfilling the two criteria of stress tolerance and high biomass are found to be effective for phytomining of areas where a high concentration of metals are immobilized, having low pH of the soil, and are preferably water soluble. However, induced phytoextraction is applied as another strategy where ethylene diamine tetra-acetic acid (EDTA), a chelating agent, is applied at the final stages of growth phase and harvested within a week or some days. This chelate-assisted metal translocation from the contaminated soil is then taken up by the plant into its part as a metal-chelate complex. After drying the metal accumulator, the crop is burnt to receive the bio ore. Anderson et al. (1998) have mentioned the use of the *B. juncea* plant as an effective hyperaccumulator of Au. Phytomining through tree species is critical in toxic pollutant removal from the soil; however, approaches in this area are still limited. Being an innovative method, this technique can be commercially promising if focused on the extraction of rare earth elements as they are high at rates and in demand. Some experiments in this area have been remarkable, through phyto-based remedial techniques to restore rare earth metals, as documented by Wu et al. (2013),

Wiche et al. (2017), Zhou et al. (2015a) and Maksimović et al. (2014). However, phytomining is possible only in mine lands with valuable metals.

Phytomining proposes mining of ores that are not feasible by conventional techniques, plant acquired metal ores or "bio ores" are less energy-consuming for restoring metals and are sulfur free. It also contains demanding metal contents compared to conventional ores and is no space constraint. *Alyssum murale* plant is a nickel hyperaccumulator.

3.2.3 Silvicultural Approach

Mining site reclamation and mitigation of its subsequent waste can be done by applying revegetation techniques (Bradshaw and Johnson 1992). Land use patterns for mining activities are found to be temporary, after which the degraded land can be revamped to develop its aesthetic appearance as well as its productivity (Maiti 2013). If these abandoned sites are remediated by developing a plant cover, much of the land can be reutilized over a while (Kramer et al. 2000; Mains et al. 2006), rather than leaving the mine site "orphan" or treating them with chemicals (del Pilar Ortega-Larrocea et al. 2010).

Resource digging from the mines is time bound as it has the limitations of being exhausted. Many decades are further required to restore these areas of excavation and return them to their original form. If this restoration is not completed successfully, the originality of the ecosystem and its sustainability are not maintained. Lima et al. (2016) have specified the "R4 terminology" which involves (i) remediation, (ii) reclamation, (iii) restoration, and (iv) rehabilitation, subject to the type of mining, technology applied, material excavated, and its effect on the environment and possible recovery activities.

A layout plan for restoration can be initiated before mining activities tend not to be fruitful any further. Being an essential step in eco-restoration, utilization and identification of trees needed for plantation are required which find its mention in most of the earlier restoration plans. It has proved to be beneficial and has been a constructive approach for opencast mining site restorations (Barnhisel and Hower 1997; Bradshaw 1997). Soil restoration is crucial as it is the prime factor for the survival of the terrestrial ecosystem. During excavation, sometimes random soil and landforms are obtained which pose challenges if remain unattended. However, suitable species of plants or a mixture of vegetation may be counted upon to address these challenges and promote afforestation in these sites. Thus, soil quality is enriched as well as biomass is added to the environment maintaining both soil and vegetation restoration through tree plantations (Lei et al. 2016).

A typical forest restoration plan on devastated lands requires some basics which can be opted as "four Ps" – Plan, Prepare, Plant, and Protect. (i) Planning the site conditions and developing forest restoration, (ii) Preparation of the site for forest cover, (iii) Plantation of mixture of valuable and indigenous tree/plant species, and (iv) Protection via management of the site for weed control, forest fire, and animal and human encroachment.

3.2.3.1 Hurdles in Phytorestoration of Mine Sites

The toxicity of different pollutants and deficiency of nutrients and organic content in the soil make it unfit for recultivation or afforestation purposes. This poses a major hurdle in revegetation of the mine sites, which causes insufficient or poor results during budding phase of afforestation (Kost et al. 1998; Holl 2002). The industrial revolution has contributed significantly to raising environmental pollution (Nriagu 1996), and anthropogenic activities release considerable metal pollutants in the soil (Alloway 1995). Affecting soil quality and leaching into ground and surface waters, they pose a threat to food chain contamination (Nowack et al. 2010). The presence of undesirable metal quantities and the deficiency of nutrient and organic matter in reclaimed soil make accumulator plant species a demanding option. Mined soil lacks carbon (C) due to decay of soil and decomposed organic matter (Ussiri and Lal 2008); as a result of reduction of above- and below-ground biomass litter. This micronutrient soil deficiency loses soil organic carbon (SOC) and disrupts the water retention capability of the soil in arid regions (Anderson et al. 2008), thereby disrupting plant cultivation. SOC not only maintains the physicochemical properties of soils but also supports the growth of the microbiome. The microorganisms are accountable for the decomposition of organic matter and taking care of the nutritional needs of the plants. In impoverished sandy soils and forest areas, nutrients stored are slowly discharged through mineral transformation (Wos and Pietrzykowski 2015). Changes in the physical and chemical environment hamper the vitality of the microorganisms associated with the soil biome (Turco et al. 1994), which may require up to 20 years to revive back to normal (Anderson et al. 2008). Though soil microbiota is vital for the survival of the terrestrial ecosystem, the major distinguishing indicators of disturbed land reclamation remain limited to soil vegetation cover and soil erosion. The soil health through evaluation of its microorganism count remains unnoticed which is undoubtedly an important criterion (Mummey et al. 2002). Mukhopadhyay et al. (2013) suggests that the well-being of coal mines depends critically on the "dehydrogenase activity, soil CO_2 flux, organic carbon, and moisture content." Thus reclamation process can be justified if the microbiota restoration and reclamation process go hand in hand, to ensure successful reclamation of the site (Mummey et al. 2002). Moreover, the tree species have the expertise in the recuperation and development of microbiotic niche in reclaimed mine soils (Chodak et al. 2015).

The major obstacle in silviculture-based recovery is the time to set up the natural habitat of the damaged site. Though it is supported that trees help in recovering the primary balance and recovering generic configuration of the assisted microbiota, Holl (2002) suggests even plant species having expedited growth can take up to 35 years to revive the entire greenery in coal mine areas. This problem can be addressed by proper management for restoration activities along with long-term planning and selection of the most appropriate plant species.

3.2.3.2 Trees in Repair

Owing to the challenges of soil and site revival, the vegetative species selection, technologies, and soil modifications are crucial. Due to the degraded soil quality as a mining impact, only stress-resistant species are preferred for the total revival of soil quality, including soil microbiota, moisture, fluid-retention capacity, nitrogen and organic carbon, particles like clay and silt, and bulk density (Dutta and Agrawal 2002). Apart from soil quality, the species density and composition of trees are equally needed for promising revegetation (Nurtjahya et al. 2008). Typical tree species provide soil quality conditioning and help in high biomass production which is beneficial for forest-based product generation such as paper, fiber, building essential, furniture, and above all energy (González-Oreja et al. 2008). An assorted collection of tree species have enhanced chances of survival, higher growth rates, and capture more carbon (Aerts and Honnay 2011). However, supplementing the soil with amendments such as manure, wood residues, and sewage sludge improves soil quality and fights nutrition deficiency.

González-Oreja et al. (2008) have pointed out certain tree species which are productive even in stressful conditions of poor soil fertility. Hence it can be ascertained that continuous afforestation programs (Prach and Hobbs 2008), if implemented on degraded mine sites can yield desirable soil quality over years.

Krzaklewski et al. (2012) indicated "three strategies for afforestation: the pioneer, the target (climax), and the biodynamic method." The combination of target-climax species with concerted agro-technological cultivation with climax species is involved in target species mechanism; however, in the biodynamic method, species with phytomelioration functions and target species, e.g. alder, are involved.

Circumstances of poor soil quality (acidic or alkaline), infertile conditions of soil, low or underrated carbon content, and compaction make revamping of such mine areas a difficult task. Planting of succession pioneer species in the beginning and later switching to other species with desired habitats can promote forest land revegetation (Pietrzykowski et al. 2015). Some pioneer woody species are listed in Table 3.3, which are capable of growth and reclamation of mine sites.

Table 3.3 Pioneer woody species in site reclamation.

Pioneer species	Attributes	Reference
Populus alba *Amorpha fruticosa*	Capable of colonizing fly ash deposits	Pavlović et al. (2004)
Robinia pseudoacacia *Alnus glutinosa*	Have the ability to fix atmospheric nitrogen, improve soil properties	Haynes (2009)
R. pseudoacacia	Nitrogen fixation, fast-growing, can tolerate pollutant, applied for post-mining landscapes	Grünewald et al. (2009), Kanzler et al. (2015)
Populus suaveolens Fisch. × *Populus trichocarpa* Torr. et Gray cv. Androscoggin)	Soil reclamation potential	Quinkenstein and Jochheim (2016)
Salix sachalinensis and *Salix viminalis* × *Salix miyabeana*	Good survival in rocky coal mines	Caterino et al. (2017)
Miscanthus giganteus	Fast-growing tree capable of growth in adverse habitat	Bang et al. (2015), Grobelak (2016)

"Short rotation coppice (SRC) systems" help gather extensive biomass through the frequent harvest of the species. Though it is mentioned that pioneer species have fast growth and provide green cover in a couple of years, the use of climax tree species provides sustainability over years. Brunori et al. (2017) have reported that *Quercus robur,* a climax species, was efficient in high CO_2 assimilation starting four years of its plantation, thereby maintaining a positive carbon balance.

It is reported that over three to five years a sulfur-rich mine site can be reclaimed by the simultaneous use of herbs, shrubs, cover crops, and successive tree plantations, which would lead to a significant increase in soil organic matter, for fruitful silviculture (Dowarah et al. 2009). Use of amendments like compost, limestone, and both in pyrite soils can help in the cultivation of *Eucalyptus grandis*, *Senna siamea*, and *Leucaena leucocephala* (Ssenku et al. 2014a). Although many plant species are effective in imparting soil reclamation through pH balance, soil texture improvement, nutrient concentration, microbiome support, and bulk density, tree species specifically, *Acacia auriculiformis*, *Eucalyptus hybrid*, and *Casuarina equisetifolia* are counted upon to be the promising species in revamping mine spoils (Maiti 2007). However, the use of herbicides may help get rid of nonbeneficial plants evading reclamation sites to ensure proper exposure to the demanding species. Thus, a mine site reclamation strategy should be developed keeping in view the suitable species, site type, time frame for restoration, and climate, which is solely restricted to experts in this area to impart effective management and control.

3.2.3.3 Benefits of Silvicultural Restoration

The silviculture technique enables the growth of trees in deserted mine areas providing an aesthetic appearance and generating a forest cover. It can be mentioned as an important step in the reclamation of large mine sites which was otherwise damaged and unproductive. The plantations of trees improve possibilities of land use such as recreation, tourism, horticulture, agriculture, and attraction to business markets. It also improves the economy of a place generating revenues from wood and wood-based products such as timber, firewood, furniture, and building types of equipment. A long-term approach in this field can impart a pollution-free environment for a sustainable ecosystem. Along with the improvement of soil quality, trees provide biomass that may be high in metal content which can be termed as waste biomass; however, these can still be utilized by applying thermal treatment in succession to reduce the metal content in further generations. They can bring economic returns if used for nonfood-based products.

3.2.4 Aided Phytostabilization

Phytomanagement (management utilizing plants) in detail phytostabilization targets a vegetation cover to promote plant growth and immobilizing trace metal elements using metal defiant plant species. Different alterations and rehabilitation through stoic plant varieties are the feature of aided phytostabilization. This technique is recommended and is a god choice as it reduces the environmental hazards and stabilizes PTM through different uptake mechanisms. Moreover, the green cover reclaims soil fertility and nutrients and restores soil microbiota over a while. In this method, pollutants are restrained within a soil periphery by active root absorption or adhesion to rhizosphere soil, arresting further migration via different channels such as water and air, preventing erosion and percolation (Bolan et al. 2011). Phytostabilization produces a green capping for the sustainability and restriction of mine slags from degraded mine sites. The plant succession in this regard reduces wind dispersion of soil; moreover, the roots arrest erosion and metal movement in the plant parts above the ground (Ernst 2005). Phytostabilization emphasizes metal absorption within root areas and restricting uptake to the exposed parts (shoot and leaves). Thus, the bioavailability of metals is limited to roots but not the shoot, reducing risks of metal contamination of livestock, wildlife, and humans through the food chain (Munshower 1994; Cunningham et al. 1995; Wong 2003).

3.2.4.1 Phytostabilization: Process and Factors

Phytostabilization, as the name suggests, utilizes plant species to stabilize or minimize metal translocation in their vegetative parts through their "extensive root system." This phenomenon needs to be adopted for mine tailings, where woody trees optimize a green cover and simultaneously restrict metal uptake in the root areas (Mendez and Maier 2008). To maintain pollutant-free soil components, restoration, and rhizosphere management, certain "plant growth promoting rhizobacteria (PGPR)" and soil organic and inorganic modifications, in addition to customized liquid permeable textile usage, can be considered. However, it should be in context to the amount of metal concentration available which can be limited by using soil modifiers (Bolan et al. 2011). A rhizosphere-mediated approach in alteration of soil organic matter, organic acids, sugars, amino acids, proteins, enzymes, etc., is reached, where microbial organic acids sometimes resemble the plant's original synthesis of the same, in terms of composition. These components then intrude the rhizosphere during cell lysis, leakage or root secretions as acetic, propionic, succinic, butyric, citric, malic, malonic, and oxalic acids to name a few (Chang et al. 2002) Besides the unavailable nutrient, trace elements such as iron, phosphorus, and other soil minerals undergo reduction and solubilization in soil, induced by these acids, and are made available to the plants (Chang et al. 2002).

Thus during heavy metal contaminated mine site restoration through phytostabilization, certain factors are highly influential, such as

i) Choice of suitable plant species which would limit metal transfer to the above-ground biomass of the plant (Yang et al. 2016a);
ii) Promote rapid vegetation prospects in stressful environments; and
iii) Improve soil fertility and microbial diversity for sustainability (Barbosa et al. 2016).

3.2.4.2 Aided Phytostabilization: An Approach to Mitigation, Stability, and Soil Modification

The use of crops at mine sites for restoration can stabilize metals by precipitation inducing chances of metal reduction and availability which will limit its adhesion to rhizosphere areas (Mirza et al. 2011). To balance the microbiota and organic matter of the soil along with N mineralization, aided phytostabilization is required (Dongmei and Changqun 2008). Reported to be a promising recurrent industrial crop, *Miscanthus* spp. is efficient in metal elimination through its active root system. It has been documented by Pavel et al. (2014) that when tested for metal removal from a metal-rich (Pb, Zn) vicinity comprising red mud as an amendment in soil, maximum accumulation of metals was limited to roots with negligible concentrations in above ground biomass. This study further appraises the use of

Miscanthus spp. as an active metal extruder plant especially lead (Pb) in damaged mine areas. Additionally use of red mud as an ameliorant (Pavel et al. 2014) to suppress metal translocation in stem and leaves can also be promoted. Such findings confirm the usage of *Miscanthus* spp. for renewable biomass generation on metal-loaded soils, as negligible metal uptake in exposed plant parts were identified. Similar results were reported in *Arundo donax* L. commonly known as giant reed, where the metal toxicity effect of Cd–Ni, was unidentified when the plant was grown in spoilt soil, rich in Cd–Ni pollutants (Papazoglou 2007).

According to Morandi et al. (2016), inputs on *Miscanthus* spp. and other crops are required which withstand stress and yield high biomass without additional requisites of fluids and nutrition. Found suitable for growing on all soil types with low water and degradation (Randall et al. 2016), *Miscanthus* spp. has high productivity in contaminated soils, and metal hoarding is comparatively more in the roots than in the vegetative parts (Barbosa et al. 2015), which make this species preferable for aided phytostabilization technique.

The species *Miscanthus sinensis* was tried with *Pteridium aquilinum* L. for aided phytostabilization potential toward mine refuses rich in Pb/Zn along with supplements of iron-rich amendments which produced a considerable reduction of extractable metals from mine refuse (Lee et al. 2014). Among the modifiers used by the scientists for the soil were slag, red mud, bone mill, bottom ash, etc., which were procured locally from power plants (coal and steel), fertilizer vendors, and alumina smelters (Lee et al. 2014). Amendments used supported in active stabilization of present toxic heavy metals and limited their uptake in test species, which promised an able stabilization strategy. Thus it can be suggested that *M. sinensis* has an effective low uptake affinity and restricts metal absorption only to its root, which nominates it to be a candidate species for aided phytostabilization compared to *P. aquilinum*; however, modifiers rich in Fe/Zn such as furnace slag and red mud are an added advantage to the process when applied in the soil (Lee et al. 2014). Use of alterations in the mine soil with applications of citric acid, EDTA, and acetic acid has also shown remarkable results in *A. donax* with volume production and restricted root metal assimilation in areas facing severe stress (Guo and Miao 2010) of As (254 mg kg^{-1}), Cd (76.1 mg kg^{-1}), and Pb (1552 mg kg^{-1}) (Miao et al. 2012). It was concluded that persisting grasses are equally efficient in adapting to a stressful environment, restricting metal absorption to their radicle. Soil quality was also enhanced by *A. donax* (L) variety which was observed in variables such as organic carbon, pH, enzyme activity, electrical conductivity, and microbial load of the soil. Apart from this, metals like cadmium, lead, cobalt, and iron remained accumulated to the belowground biomass when planted in an area under metal stress supplemented with red mud derived from bauxite ore (Alshaal et al. 2013).

To supplement mine soils with organic matter as it is deficient in it, the use of biochar, agricultural residues, industrial organic refuses, animal dung, and woodchips along with garbage or solid wastes can be done (Li and Huang 2015). The use of pyrolyzed biomass (biochar) as a modifier in aided phytostabilization was further advocated by Fellet et al. (2011). Microbial count of the soil was also enhanced through phytostabilization, also the use of *Silphium perfoliatum* L. promoted soil biodiversity and ecosystem restoration through a time frame of nine years by imparting an earthworm niche (anecic and endogeic) in marginal lands. Earthworms in soil also added to the increment of N_2O and CO_2 (Lubbers et al. 2015). Soil amendments, and microbiota if combined with root discharges, can provide a new tailored soil in place of mine wastes. The use of perennial grasses also initiates stimulation of microbiota adding to the assembly of C and N in soil (Cattaneo et al. 2014). This combined application can be cheap, effective, eco-friendly, and sustainable in pollution reduction (Nsanganwimana et al. 2014). As nitrogen remains an undoubted requirement in mine soils, alternatively certain legumes or nitrogen fixers can also be introduced in these areas, or application of nitrogen-based fertilizer may be an option (Li 2006).

Use of industrial crops on wastelands not only impart soil quality and fertility, control erosion and leaching, and arrest pollution of groundwaters but also help in gifting an improved topography, countryside expansion, revenue generation, energy supply, and reduction in greenhouse gases (GHGs) (Fernando et al. 2010). For plantations in sloppy terrains, giant reed is desirable as it has a fighting fit nature preferable for damaged soils (Fernando et al. 2018) which will control soil loss and evapotranspiration supporting plant growth (Yang et al. 2014).

Leaching control of contaminants during AMD is boosted by phytostabilization, plant species capable of filtering ions from wastewaters can be established to arrest metal contamination risk to an extent. In this process, some perennial plant species can be utilized to isolate nutrients from effluent (Costa et al. 2016). A 10-week study carried out by scientists to mark the level of leachate for Zn, Cd, and Cu from effluent using *Phalaris arundinacea* and *Chamaenerion angustifolium* species (Neuschütz and Greger 2009) gave assuring results of lowered concentrations and less leachate material when compared to control. Thus, pioneering *P. arundinacea* on quarry wastelands along with slime can be propitious for phytostabilization, although caustic supplements are essential to balance pH fluctuations. Apart from these, certain other native plant varieties belonging to different families have also been documented after study in different site areas (Table 3.4). These are based on individual researches documented in articles in the context to metal assimilation in plant aerial parts not exceeding household animal toxicity limits (NRC 2005).

However, in the long term, the final destiny of metals in afforested mine lands remains uncovered and needs investigation. This is especially required to estimate

Table 3.4 Some potential plant varieties in aided phytostabilization.

Plant	Family	Metal(s) identified	Study area	Reference
Atriplex lentiformis	Amaranthaceae	As, Cu, Mn, Pb, Zn	USA	Mendez et al. (2007)
Baccharis neglecta	Asteraceae	As	Mexico	Flores-Tavizon et al. (2003)
Bidens humilis	Asteraceae	Ag, As, Cd, Cu, Pb, Zn	Ecuador	Bech et al. (2002)
Cymbopogon citratus	Poaceae	Cr	India	Kumar and Maiti (2015)
Euphorbia cheiradenia	Euphorbiaceae	As	Iran	Malayeri et al. (2013)
Jatropha curcas	Euphorbiaceae	Zn, Pb, Cu	Philippines	Aggangan et al. (2017)
Lygeum spartum	Poaceae	Cu, Pb, Zn	Spain	Conesa et al. (2007)
Typha latifolia	Typhaceae	Cu, Zn, Mn, Pb, Ni	India	Maiti and Jaiswal 2008
Sapium sebiferum	Euphorbiaceae	Pb, Zn	China	Tang et al. (2019)

the virtue of phytostabilization in persistent metal removal, benefiting soil attributes, and revegetation. It is already acclaimed that phytostabilization has a promising future in dry and deserted areas; however, an ecosystem comparable to natural grasslands is a bit challenging and may require many years. The fact that phytostabilization caps specific metal uptake in plants is, however, limited to plants having metal affinity and are metal accumulator species. As the metals are restricted in the lower part of the plant (roots) long-term efforts of plant successions can help alleviate metal pollution converting degraded mine lands to lush green landscapes.

3.2.5 Microorganisms in Remediation

Over centuries of mining activities, the Earth has faced tremendous impact from deforestation and piling of mining wastes. Industrialization and its demand have unearthed almost more than 50% of underground reserves of coal and rare earth metals. Due to this, the quality and environment of the areas have changed drastically and have been overburdened by wastes that require immediate attention. As the environment is getting polluted, there appears a change in the microbial diversity, hampering the soil quality and fertility. Over a while, this has resulted in the loss of vegetation as diverse microorganisms in soil inevitably support the development of plants. Post mining, the orphan mine sites lay unattended and contaminated with enormous wastes causing health hazards in multifold. Certain tangible and synthetic methods have been relied upon to reduce the lethality which comes at a cost and lacks labor skills; thus, the restoration of these mining sites seems a burden and remains finite (Tsydenova and Bengtsson 2011; Förstner and Wittmann 2012; Dhal et al. 2013; Sharma and Pandey 2014). Recently, revegetation and phytoremediation methods are involved in metal-loaded sites (Sheoran et al. 2010; Nirola et al. 2016; Venkateswarlu et al. 2016), and extra stress should be laid in determining the function of microbiota and plant relationship in mining sites to demonstrate a successful ecorestoration.

Microorganisms have a major role in the reclamation of mine sites (Singh et al. 2011; Li and Zhang 2012; Ma et al. 2016). Their natural tendency to form colonies and adaptability in devastated lands help generate a new ecosystem. Living organisms such as algae, herbs, and plants including shrubs and trees are benefitted immensely as the unicellular microbes alter soil chemistry establishing an apt environment for their growth. During studies of relevant kinds of literature related to microbes, it was noticed that the presence of certain fungal species has initiated efficient metal assimilation in plants by supplying added nutritive value (Sheoran et al. 2010; Miransari 2011; Meier et al. 2012a). Plant growth promoting (PGP) bacteria and fungus are utilized in practice for bioremediation techniques; but the on-site application of these is somewhat not very feasible on large areas (Mench et al. 2010; Solís-Domínguez et al. 2011; Kleczewski et al. 2012). Among the fungal species, mycorrhizal fungi are utilized in mine site rehabilitation compared to endophytic fungi (Khan et al. 2000; Turnau and Haselwandter 2002; Jeffries et al. 2003; Gaur and Adholeya 2004; Jasper 2007; Reverchon et al. 2015).

Microorganisms in PGP are of immense use due to their growth-promoting traits such as growth hormones and siderophores, useful for the quality of the soil and enhanced plant growth (Dell'Amico et al. 2005; Grandlic et al. 2008; Berg 2009; Ma et al. 2011; Kumar et al. 2017). This not only minimizes agricultural costs by cutting the use of synthetic fertilizers but also improves soil quality and pollutant resistance and manages soil microflora (Jiang et al. 2008; Upadhyay et al. 2009; Ahemad and Khan 2011; Becerra-Castro et al. 2012; Ghnaya et al. 2015; Tripti et al. 2015). Different variables of fungi, such as rhizospheric soil and endophytic fungi, have been important in the reclamation of a mine site, which supports that fungi irrespective of their kind is beneficial for the restoration of damaged soils. It is nowadays in practice to club both fungi and PGPR for restoration purposes as they [ectomycorrhizal (EM) fungi and rhizobacteria] are found in abundance in the rhizosphere responsible for efficient metal accumulation and regulation of geochemical cycles. Bacteria and mycorrhizal fungi can be used as "natural biofilters" for supplementing nutrition and metal accumulation in plants, as documented by Haferburg and Kothe (2010). These microbes blend well with disturbed soil and impart efficient remediation through plants (Zimmer et al. 2009). However, the abundance and concentration of fungi and bacteria count upon certain components such as nutritional value of the soil, plant species, and environment. However, scientists have reported more than 80% of terrestrial plants showing the reciprocal relationship with arbuscular mycorrhizal fungi (AMF), which manages their growth in devastated lands (Redecker et al. 2000; Jeffries et al. 2003; Khan 2005; Finlay 2008; Bonfante and Anca 2009). It has been documented that most of the fungal communities are active in different layers of soil such as topsoil, mine leachate, mine refuse, root zone, and phyllosphere as endo- or epiphytes.

3.2.5.1 Metal Tolerance in Mycorrhiza Fungi

Biosorption is a distinct phenomenon of metal absorption, which enables microorganisms to bind and impound metals in the soil. Many metal-containing discharges can be treated by a microbial community which is efficient as well as cost effective. Malik (2004) and Tobin (2001) have termed yeast and fungi as mycosorbents capable of mycosorption, as fungi have gained much recognition in metal accumulation and recovery suitable for mine sites. Metal uptake by fungi can be distributed into biosorption, fungal intracellular uptake either by dead or living, and metal alteration by live fungi. Nonliving fungal matter has metal affinity during metabolism as proton production is negative and depends on precipitation, crystallization, chelation, and ion exchange properties for biosorption (Danesh et al. 2013). Plant with required nutrients (Hodge and Fitter 2010; Miransari 2010; Turnau et al. 2010; Vafadar et al. 2014; Nadeem et al. 2014).

Mycorrhiza is omnipresent, plus well suited for burdened environments, and can readily accept and acclimatize with natural and artificial biospheres. These make hassle-free union with almost all plants and aid in plant well-being such as growth by associating its mycelium in the roots and imparting nutrients; as a biodegrader (destroys pathogenic microbes by producing secondary metabolites (Pineda et al. 2010; Mendes et al. 2013; Pusztahelyi et al. 2015), drought resistance through water-retaining capability and metal absorption by expanding root structure and surface area, and above all improves soil microbial diversity by increasing soil carbon required for microbial existence (Rillig 2004; Herman et al. 2012; Asmelash et al. 2016). Certain fungi and associated plant species used in mine site reclamation globally are listed in Table 3.5.

Increased photosynthetic and transpiratory activities are noticed in mycorrhizal plants that focus on the development and metabolism of AMF (Augé et al. 2015). This depends on the transportation of water and nutrients in the plant which is favored by mycorrhization.

Soil nutrient content is hampered due to toxic metals, and it is then that the plants secrete organic acids which bind with the metals (Sharma and Dietz 2009) and reduce toxicity to the plant, AMF, and other surrounding microbiota. Thus

Table 3.5 Fungi and plants used in mine site rehabilitation.

Fungal species	Plant studied	References
Facultative arbuscular mycorrhizal fungi	*Molinia caerulea*	Bąba et al. (2016)
Rhizophagus, *Glomus*, *Funneliformis*, *Acaulospora*, *Diversispora*, *Claroideoglomus*, *Scutellopora*, *Gigaspora*, *Ambispora*, *Praglomus*, and *Archaeospora*	*Digitaria violascens*, *Eclipta prostrata*, *Veronica didyma Tenore*, *Digitaria ciliaris*, *Herba seu Radix Amaranthi*, *Eleusine indica*, *Solanum nigrum*, *Miscanthus floridulus*, *Allium senescens*, *Clinopodium chinense*, *Poa annua*, and *Paspalum paspaloides*	Sun et al. (2016)
Rhizophagus clarus, *Gigaspora decipiens*, and *Scutellospora* sp.	*Albizia saman* and *Paraserianthes falcataria*	Wulandari et al. (2016)
Funneliformis mosseae and *Funneliformis geosporum*	*Miscanthus* × *giganteus*	Firmin et al. (2015)
Glomus intraradices	*Zea mays* and *Plantago lanceolata*	Turnau et al. (2012)
Glomus, *Gigaspora*, *Acaulospora*, *Entrophospora*, and *Sclerocystis*	*Dalbergia sissoo*, *Acacia auriculiformis*, and *Cassia siamea*	Mukhopadhyay and Maiti (2011)
G. intraradices and *Glomus deserticola*	*Prosopis juliflora*	Solís- Domínguez et al. (2011)

the plant's root exudates function in increased root AMF symbiosis to immobilize and extract metals from the sites.

3.2.5.2 Application of AMF in Revegetation and Remediation

AMF are common in most terrestrial plants. Most revegetation strategies remain dependent on the nature and existence of AMF on the plant species, where fewer agents of reproduction such as hyphae, spores, and colonized roots and low AMF give limited or less productive results of vegetation in the preliminary stages of reclamation. Thus soil management has to be carried out where reproductive propagules are introduced in native plant species and can be pioneered. However, the AMF species can be considered based on the earlier studies conducted on similar areas of contamination. Leung et al. (2010b) and Chen et al. (2006) have highlighted that metal accumulator species that are most efficient in metal uptake in association with plants are best to be adopted for remediation purposes.

A similar finding was reported by Bhalerao (2013) that isolate selection is critical based on its tolerance toward metal contamination as well as assuring the plant's survival in the contaminated site (Bhalerao 2013).

On identifying such variants helpful in host plant growth along with metal removal from the sites, the reproductive portions can be consolidated and introduced in areas with less developed AMF colonies in native plants. AMFs can be cultured in vitro or as pot-based microcosm study with plants adopting excess concentration of fungi (Diop 2003; Gaur and Varma 2007). Increased tolerance toward heavy metals (HMs) and soil remediation are the important aspects of phytoremediation studies with *increased extraction of metals reported in* isolate species of *Glomus mosseae* (Cd, Zn, and Cu), *Glomus claroideum* (Cu and Zn), *Glomus intraradices* (Ni, Zn, and Pb), and *Glomus margarita* species (Joner et al. 2000; Tonin et al. 2001; Gonzalez-Chavez et al. 2002a; Al Agely et al. 2005; Marques et al. 2006; Turnau et al. 2008; Orlowska et al. 2012; Yang et al. 2016a).

3.2.5.3 Issues and Prospects

For proper reclamation of degraded lands, an integrated treatment with plant and AMF must be adopted. This will not only help in restoring mine sites but also improve soil quality (Tisdall 1994; Borie et al. 2008) and microbial diversity to restore nutrient cycling. Being an important ingredient in phytoextraction or phytostabilization, they make the survival of plants easier in otherwise compromised areas (Leung et al. 2010a). The metal-holding feature of the hyphae of AMFs has been recorded (Gonzalez-Chavez et al. 2002a, b; Cabral et al. 2010), and laboratory-based experiments too have proved the role of AMF in phytostabilization of HMs (Leyval et al. 2002; Rufyikiri et al. 2002). However, the exact role and integrated approach of the AMF species are not completely known, and soil bioremediation potential is yet to be determined. Long-term approaches of field application may provide next-level blueprint for biotechnological implementations.

3.2.6 PGPR-Assisted Phytoremediation

Post mining, the "orphan" mines contribute immensely in piling up heaps of contaminants causing environmental concerns. Wastes include mine tailings, slags, and AMD which cause land and water pollution because of metals that are hazardous for living creatures. Mine site reclamation becomes challenging as the discarded waste, and conventional techniques to manage them become impractical and expensive (Ali et al. 2013; Thijs et al. 2017). The strategy named phytoremediation has gained importance over some decades and has also delivered promising results on application. Being cheap and eco-friendly and having assured results, this has negligible side effects on the ecosystem. Metal pollution in soil has now been in studies, and the role of soil microbes is also been acknowledged in terms of metal uptake and reaction toward environmental stress. Phytoremediation assisted with indigenous microbes of the soil, commonly termed as rhizobacteria capable of plant growth promotion, have commendable ability toward amplified development and metal assimilation by the plants through alterations in the solubility, chelation, and pH regulations (Ma et al. 2011). Microbial diversity participates in crop productivity, and PGPR has shown quite effective results in this regard, by triggering plant growth through related or indirect modes like phytohormones, bioactive substance release, nutrient accessibility, and absorption along with protection against soil-borne pathogenic microbes. PGPR thus plays a pivotal role in soil fertility management (Bishnoi 2015).

3.2.6.1 Rhizosphere: The PGPR Territory

The rhizosphere area is the soil adjacent to the plant root system which is intricate and rich in bacteria, fungi, and numerous other microbiota (Venturi and Keel 2016). The bacterial species in this area are termed rhizobacteria and are rich in species diversity, as the nutrient-rich zone provides sugars and ample amino acids that serve as a source of constant energy, derived out of the plant exudates (Badri et al. 2009; Beneduzi et al. 2012; Goswami et al. 2016).

Huang et al. (2014) mentioned that whether these rhizobacteria are fruitful, noxious, or indifferent is directly proportional to their effect on the respective plants. Rhizobacteria are attracted by the root released enzymes which are high in nitrogen and carbon contents, are converted into absorbable molecules which are gained by the roots again, and return to promote overall health of the plants. Some of the common organic compounds identified in root exudates are organic acids such as citric, malonic, glutaric, amino acid and enzymes such as invertase, phosphatase, amylase, phenolic acid, coumarin, and sugars.

3.2.6.2 Plant Growth Promoting Rhizobacteria

Rhizobacteria beneficial for plant growth and nutrient uptake were coined by Kloepper and Schroth (1978), as PGPR, which in plant association provide fruitful results in terms of production. PGPR strains work in numerous ways such as biofertilizers, biopesticides, and phytostimulators (Vessey 2003; Somers et al. 2004; Chandler et al. 2008). Soil microbiotas in bulk and rhizosphere zone have disparities as the microbiome near the root vicinity is in abundance compared to the other.

Different root exudates alter the microbial colony near the root zone concerning the plant demands. Thus, a niche for typical plant–microbe interaction is created for specific plant species (Broeckling et al. 2008; Berg and Smalla 2009; Berendsen et al. 2012).

Further Gray and Smith (2005) have subclassified the types of existing PGPRs based on their existence and role into extracellular PGPR (ePGPR) and iPGPR. ePGPR exists in the surroundings of the root or the intercellular vacuoles of the cortex, whereas the latter resides inside the root cells in nodular forms. Some of ePGPR are *Micrococcus*, *Chromobacterium*, *Agrobacterium*, *Caulobacter*, *Flavobacterium Arthrobacter*, *Pseudomonas*, *Azotobacter*, *Bacillus*, *Erwinia*, *Burkholderia*, *Azospirillum*, and *Serratia* (Gray and Smith 2005), while the examples of iPGPR are *Rhizobium Allorhizobium*, *Bradyrhizobium*, *Azorhizobium*, *Mesorhizobium*, and of the family Rhizobiaceae and *Frankia* species (Verma et al. 2010). PGPR strains participate in plant interests by two methods: direct and indirect. The direct method includes PGPR as biofertilizer by biological nitrogen fixation (BNF), iron sequestration, production of certain phytohormones such as IAA, gibberellin, auxin, cytokinin, and phosphorus solubilization. Indirect methods include biocontrol as a biopesticide, bioremediation, and assisted phytoremediation.

3.2.6.3 PGPR as Biofertilizer

A basic requirement for the synthesis of all proteins is nitrogen. Though it is present in abundance in the environment, it is not easily available to the growing plants due to its chemically inactive nature. However, the atmospheric nitrogen is utilized by nitrogen-fixing microorganisms by a method termed BNF by using the enzyme nitrogenase. This process involves symbiosis of leguminous roots of plants with the nitrogen-fixing bacteria and helps restore the nitrogen requirement of the plants. Some varieties of nitrogen-fixing bacteria helping in plant growth and efficiency are *Sinorhizobium, Bradyrhizobium, Rhizobium,* and *Azorhizobium,* collectively termed rhizobia. This BNF supports a high yield of vegetation over large landmasses and relatively cuts the cost of expensive nitrogenous fertilizers. Such microbes when incorporated in damaged soils can help reclamation of large areas in a coherent manner.

Iron is the most crucial cofactor for most hormonal and enzymatic reactions, exists in the form of complex oxides (Fe^{3+}) which are not readily available either to the plant or the bacterial communities. However, certain bacterial strains produce less molecular weight compounds named siderophores having great iron-binding affinity in iron-deficient media and help in forming a Fe-siderophore complex (Glick 2012) which makes iron readily available to the plants.

Phytohormones are complex compounds which in fewer amounts impact and modulate the metabolic mechanisms in plants (Fuentes-Ramirez and Caballero-Mellado 2005). These PGPR strains are capable of certain essential phytohormones such as indole-3-acetic acid (IAA), gibberellic acid (GA), and cytokinins or by ethylene level regulations in plants (Glick et al. 1998; Bottini et al. 2004; Spaepen et al. 2007).

It has been earlier documented that most of the microorganisms extracted from the rhizospheres can release and produce auxins as IAA which acts as a signaling molecule in the plant–rhizobacteria symbiosis. IAA is essential in cell division and differentiation; initiates seed and tuber germination; induces effective rapid root development; optimizes vegetative growth; modulates root formation and photosynthesis; and initiates responses of metabolites toward stress factors, by impeding the auxin concentration of the plant. Scientists have supported *Azospirillum* to be the most effective IAA producer among several other PGPR species (Dobbelaere et al. 1999). Others commonly include *Pseudomonas, Azotobacter, Bacillus, Aeromonas, Enterobacter, Burkholderia,* and *Rhizobium* genera, identified from separate rhizosphere zones (Spaepen et al. 2007).

Another phytohormone entailed in all-round development and plant well-being are the Gibberellins GA. Although their major role is in support of root elongation, mostly GA is effective in combination with other relevant hormones which are efficiently produced by certain PGPR strains that are studied earlier such as *Bacillus* spp., *Gluconobacter diazotrophicus, Acinetobacter calcoaceticus, Herbaspirillum seropedicae, Achromobacter xylosoxidans, Azotobacter* spp., and *rhizobia* (Bottini et al. 2004; Dodd et al. 2010). However, the precise operation of prompting plant growth by GA is quite undefined.

In plant root, meristem differentiation, cell division, and root hair proliferation are promoted by cytokinin, apart from root elongation and lateral root formation which is not supported by it (Riefler et al. 2006). PGPRs producing cytokinins have essential importance in plant root nodulation which is induced by symbiotic N_2-fixing bacteria, mycorrhizal fungus *Rhizopogon roseulus* free-living N_2-fixing bacteria, *Rhizobium*, *Azotobacter*, and *Azospirillum*. Moreover, cytokinin in the form of zeatin has been evident in PGPR-like *Bradyrhizobium japonicum, Azospirillum brasilense, Bacillus licheniformis, Pseudomonas fluorescens, Arthrobacter giacomelloi,* and *Paenibacillus polymyxa* (Timmusk et al. 1999; Perrig et al. 2007).

Phosphorus solubilization is another effective approach of the PGPR strains. Phosphorus is the next important macronutrient succeeding nitrogen required for magnification and vigor. However, the availability of phosphates to plants is critical as it occurs in two major forms: mineral and organic. Mineral forms include apatites (oxy and hydroxyl), whereas inositol phosphate (soil phytate) and phosphate esters (mono, di, and tri) make up the organic forms. Globally, P availability in soil is a major concern as far as crop yield is measured, as almost 30–40% deficit in agriculture is recorded worldwide due to reduced P levels in cultivable lands (Vance et al. 2003). In soil biogeochemistry, there is a major function of phosphate mineralization by phosphate-solubilizing bacteria (PSB) (Richardson 2001; Jeffries et al. 2003), along with plant promotion by PGPR (Rodriguez et al. 2006). Certain traits such as *Microbacterium, Azotobacter, Enterobacter, Bacillus, Pseudomonas, Rhizobium,* and *Serratia* are reported as important PSB (Bhattacharyya and Jha 2012). Other crucial strains reported to be effective as PSBs in edible crops are *B. japonicum, Azotobacter chroococcum, Enterobacter agglomerans*, *Bacillus circulans*, *Pseudomonas chlororaphis*, *Cladosporium herbarum*, and *Rhizobium leguminosarum* (Antoun et al. 1998; Chabot et al. 1998; Cattelan et al. 1999).

Initiating BNF along with P solubilization for enhanced growth of plants is an added quality of PS strains along with furnishing of trace element uptake. Thus, supplementing the crops or soil with P solubilizing strains belonging to the PGPRs may prove to be quite favorable in the retrieval of inorganic phosphates from the medium rather than an external application of phosphate fertilizers.

3.2.6.4 PGPR in Biocontrol

Plant diseases are common, and every year huge depletion in crop yield are incurred due to it (Lugtenberg et al. 2002). Undue costs of pesticides can be curtailed if PGPR strains are applied as alternatives in pest control to protect crops. PGPR strains possess the caliber of producing antibiotics, HCN and siderophores, and bacteriocins along with certain enzymes including chitinase, proteases, and lipases, which are familiar biocontrol agents against phytopathogenic escalations (Lugtenberg and Kamilova 2009). Interestingly *A. brasilense* cells have been found containing a low molecular weight growth inhibitor against a weed parasite of sunflower named *Orobanche aegyptiaca* (Dadon et al. 2004). Similarly, tropical sorghum infecting parasitic striga weed seed (*Striga hermonthica*) was inhibited by *Azospirillum* spp. supporting sorghum growth (Bouillant et al. 1997). Thus, use of PGPR is not constrained to specific group of pathogens and has crossed boundaries to prove their potency against pathogens such as bacterial, viral, and nematodes (Kloepper et al. 2004).

The function of PGPR as biocontrol representative can be pledged for the future with the emergence of sophisticated biotechnological tools to deliver more promising bioproducts in subsequent time (Berg et al. 2013). On a similar note, PGPRs capable of breaking down organic contaminant, such as polychlorinated biphenyls (PCBs), derived from degraded sites have been reported to consist of encoding genes for the same by Brazil et al. (1995). Explosives, herbicides, or hydrocarbons degrading endophytic bacteria have also been documented from similar sites (Germaine et al. 2006), along with their tolerance toward toxic metals, inorganic and organic compounds such as trichloroethylene, or polyaromatic hydrocarbon, such as benzene, toluene, etc. (Moore et al. 2006; Doty 2008). A reasonable blended approach using PGPRs and metal-degrading bacteria to refine degradation efficiency and metal tolerance by bacterial strains can be implemented to mitigate polluted site contamination (Huang et al. 2005).

PGPRs with the ability of specific or multiple metal resistance have already been studied by scientists where uptake of Nickel, arsenic, cadmium, magnesium, zinc, and copper has been documented (Abou-Shanab et al. 2006). Thus, a minute study of the adopted mechanisms by the PGPR is obligatory to enable their total ability in agronomy along with required technologies for understanding the integrated plant–microbe interaction would ease their practice in arable lands.

3.2.7 Transgenic Plants and CRISPR

Over the years, repeated dumping of mine slags and shortage of legalities have unprecedented effects on our environment, which ultimately leads to pollution of soil, water, and arable lands. This is responsible for the metal translocation through food chains to various trophic levels impacting human health. Hence, enzymatic conversion or intracellular accumulation into less toxic elements through degradation is required. Moreover, the lack of floral species for biochemical cycling and nutrient transformation is a constraint. Thus, native and genetically modeled microorganisms can be promising to clean up the toxic pollutants from mine refuse in an environmentally safe ambiance. Plant's ability in metal sequestration is long known. Most of these include *Viola calaminaria, Thlaspi* species, *Hau-maniastrum katangense, Silene cucubalus, Sebertia acuminata,* and others (Baker et al. 1988) These particular plant species provide detailed guidance on the mechanisms and approaches toward metal/s uptake and sequestration, thus using cell and engineering tools more such species tolerant towards metals can be introduced. As metals uptake in animals is arbitrated by metallothioneins (MTs) (Reddy and Prasad 1990), the CaMV35S promoter initiates a human *metallothionein* gene for plant gene expression as drafted by Misra and Gedamu (1989). Such construct was introduced to produce transgenic plants with variable metallothionein, which exhibited upgraded tolerance toward cadmium when grown in spiked soil in testing grounds. Alternatively, the use of somaclonal variations to generate metal-resistant plant cell lines can be done, where cell culture would include metal tolerance testing with a continuous increase in metal ions. Repeated testing can be followed to ensure the successful production of metal tolerant varieties of plants (Collins and Petolino 2012). The efficiency of plants in uptake and transformation of aromatic toxic compounds is also earlier acclaimed (Facchetti et al. 1986; Aranda et al. 1989; Scheunert et al. 1989; Topp et al. 1989; Bellin and O'Connor 1990; Wild et al. 1992).

Plants bear all the essential characteristics and qualities to prove themselves as the inevitable decision for the reclamation of debased mine conditions. The advancement of new gene-based manipulation tools and superior vegetation quality have accelerated progress in designing methods for better removal of toxins. Phytoextraction system is satisfactory for tolerably sullied soils just as mine lands. Around the world, more than 520 hyperaccumulators have been recognized as having *Brassicaceae* plant varieties and *Euphorbiaceae* in various polluted zones. Such metallophytes are hugely utilized for biochemical and geochemical investigation.

Metal uptake, translocation, and decontamination mechanisms in woody and herbaceous plants are similar in terms of their physiological and molecular pathways. This is useful particularly for the production of high biomass in woody plants. Due to their high phytoextraction ability and economic utility, the species of *Salix* and *Populus* are duly endorsed as the most promising plants for phyto-based remediation. The long roots of this species provide phytostabilization of metal contaminants; also the long flowering time weakens genetic spreading. Though it does not add up in the food chain, its extensive root system is useful for phytostabilization. Used as a suitable "version in forest genetics" (Tuskan et al. 2006), *Populus trichocarpa* genome sequence was employed for extensive and detailed study. The genome sequences of *Populus* and *Salix* show considerable similarities; however, the *Populus* genome is fourfold of *Arabidopsis* and lesser than *Pinus* (Marmiroli et al. 2011), whereas the haploid genome of *Salix* spp. is 500 Mbp and consisting of 19 chromosomes. Thus, a detailed study of *Populus* can help in optimizing the *Salix* genome too.

Different components, such as availability of metals in the contaminated areas, soil quality, span of trees, and unsystematic seasons, are responsible for metal uptake in woody trees. Along with this, various metabolomics tools for assessing the molecular pathways involved in metal accumulation and decontamination are also needed. During recent researches, the genetic model required in metal equilibrium is bounded to *Populus* (Marmiroli et al. 2011; Luo et al. 2016).

Being a model plant of the herb variety, *Arabidopsis* has also gained importance in terms of molecular mechanisms of metal transport over the years. Due to the intricate heavy metal balance in woody plants, *Arabidopsis* and rice were comparatively less than 70 metal transporter genes in *P. trichocarpa* (Migeon et al. 2010).

Majority of the transporters are from the natural resistance-associated macrophage protein (NRAMP), the zinc-regulated transporter, Yellow stripe 1-like (YSL) subfamily of the oligopeptide transporter (OPT), P1B-type subfamily of P-type ATPases (also known as heavy metal ATPases, HMAs), and multidrug and toxin efflux (MATE) protein family. The transporter assists in metal accumulation for which the accumulator plants are engaged through genetic engineering tools (Yang et al. 2015). Grotz and Guerinot (2006) identified *ZIP* transporters, and Adams et al. (2012) found a *Populus* genome, where about 13 ZIP families are included.

In several plants, *ZIP* transporters and the *Populus* genome are highly identified (Grotz and Guerinot 2006; Adams et al. 2012). Increased plasma membrane proton ATPases, mRNA levels of metal transporter genes, *ABCC10, ZIP2, ZIP6.2, NRAMP1.1*, and *NRAMP1.3* were responsible for cadmium influx in the lower plant biomass of the hybrid *Populus* plants (Ma et al. 2014; He et al. 2015). However, an upgrade in the level of mRNA of *ZIP6.2*, *ZIP2*, and *ZIP7.2* in the roots of *Populus* was identified under conditions of enhanced Zn addition (Shi et al. 2015).

3.2.7.1 Root–Shoot Metal Transport

In woody plants, symplastic pathways through metal carriers are initiated for the transfer of metals from the cytoplasm to aerial parts. The movement of heavy metals through the plasma membrane is mainly due to ATPase known as P_{1B}-type. Genome analysis and the qRT-PCR determined metal stress in *Populus* due to the presence of *heavy metal ATPase (HMA)* gene (Verret et al. 2004) which was evident through the presence of 17 *HMA* genes in *P. trichocarpa* (Mills et al. 2003). Different expression patterns in the plant parts were observed under variable metal load in the case of *HMA* genes in *Populus* (Li et al. 2015). The cadmium transportation pattern via plant parts of *Populus tremula* × *Populus alba* was done with the use of Cd-dithizone staining. Later, various genechip studies documented disparate regulation of genes in Cd manifested bark tissues of *Populus* (He et al. 2013).

3.2.7.2 Metal Sequestration and Metal Chelation

Although limited knowledge about the heavy metal uptake and detoxification in woody plants through proteomic and transcription regulation is available, heavy metal discharge from the cytoplasm into the organelles limited to vacuolar sequestration and the role of CDFs or MTP in this process are studied (Verbruggen et al. 2009). Some chelators have been identified in woody plant species against the herbaceous ones. Metal chelation is possible with lighter ligands such as nicotinamide and organic acids which promote metal uptake from the base to the top of the plant. Peptides such as phytochelatins (PCs) find importance in the cytosol and specific intracellular compartments along with MTs. Thangavel et al. (2007) have reported the use of PCs as early markers in metal stress detection in the cell suspension cultures *Picea rubens*. ATP-binding cassette (ABC) transporters are involved in the transportation of Metal-PC complexes. In *Rauvolfia serpentina* and *P. rubens* cell cultures, Cd ions stimulated PC synthesis have been reported (Hassinen et al. 2011). Cadmium assimilation in root, phloem, and mesophyll cells of *Populus* was studied, which when correlated with concentrations of sulfur confirmed that sulfur components have assisted in Cd sequestration. Increased metal tolerance in plants is also ensured through modification of thiols and relevant biosynthetic intermediates. However, more Cd resistance and maximum accumulation of Cd in the vegetative parts were visible in the hydroponic variety of transgenic *Populus* in contrast to the wild type (Koprivova et al. 2002; Ivanova et al. 2011).

Limited information is available in context to the physiological and molecular process which links with metal accumulation and resistance in *Populus* and *Salix* species which needs more research and exploration.

3.2.7.3 CRISPR "omics" Approaches

Populus and *Salix are* the two species to which gene involvement and molecular pathways in heavy metal balance in the hardy plants are constrained. However, with "omics" technologies (Maestri and Marmiroli 2011) such as transcriptomics, genomics, and proteomics (Jan et al. 2016), the relevant gap about genes and their complex mechanisms can be curtailed. Looking at the molecular level, scientific knowledge of RNA-Seq and/or digital gene expression are highly recommended for elaborate mechanisms, compared to the conventional gene-replication approaches. Studying the plants without genomic information (Wang et al. 2009) through RNA-Seq identification of unique genes (Mardis 2017) have proved to be quite successful in this regard. Further studies related to sequence analysis on *Salix* and *Populus* (Zou et al. 2017) have identified some new genes involved in metal resistance and uptake (Shi et al. 2016; Zhang et al. 2017). Further, in *Platanus acerifolia* tree under Pb stress, degradome sequencing also referred to as parallel analysis of RNA ends (PARE) was used to identify the microRNAs and their target genes (Wang et al. 2015).

In case of genome editing, mainly three nucleases: (i) clustered regularly interspaced short palindrome repeats (CRISPR), (ii) zinc-finger nuclease (ZFN), and (iii) transcription activator-like effector nuclease (TALEN), and are counted upon. However, using CRISPR/Cas9 is approachable for gene modelling of life forms, including plants. For efficient phytoremediation strategies of woody plants, this technique is quite obscure for uplifting the metal remediation potential of the plants through genome edition. Proper understanding of the gene-related mechanisms and transcription regulation in woody plants including *Populus* is important as large-scale mutant resources are insufficient (Fladung and Polak 2012).

In *Populus,* CRISPR/Cas9-mediated genome editing (Fan et al. 2015) was targeted by the 4-coumarate: CoA ligase gene family responsible for phenylpropanoid metabolism (hybrid *Populus*) (Zhou et al. 2015b) and located an inherent phytoene desaturase gene (*PtoPDS*) (*Populus tomentosa*) (Fan et al. 2015). Instead of being a powerful aid for gene engineering, the CRISPR/Cas9 system is target site dependent, Cas9 action, delivery method, gRNA design, etc. (Peng et al. 2016). Scientifically modified mechanisms for gene edition like CRISPR/Cas9 in plants are efficient in drawing plant genomes utilization for sustainable environmental restoration through phytoremediation. However, this technology is unfamiliar for the improvement of plant's remedial activities but can deliver reliable properties for enhancing metal uptake in plants through elaborate research.

3.2.8 Engineered Soils Using Amendments

Mining and quarrying are prevalent for ages (Tiwary 2001; Lottermoser 2010; Neri and Sánchez 2010) which include coal, ore, and rocks and mineral extraction.

Uncontrolled events of mining activities have led to environmental impacts, which is due to the irresponsible behavior of the industrial companies. Due to this, landscape deterioration, soil degradation, water and air contamination, and biodiversity loss have emanated globally (Wong 2003; Clemente et al. 2004; FAO and ITPS 2015). Surface mining removes original soil and host rocks are left as remains, due to which the soil quality, moisture, organic content, and native microbes efficient in biogeochemical cycling are lost giving a desolate look. Adding to the problem the drainage water having poor quality metallic wastes and other hazardous elements, also alters the adjacent soil chemistry endangering biodiversity and ecosystem (ELAW 2010; Lottermoser 2010; Zhengfu et al. 2010).

3.2.8.1 Soil Remediation Technologies

Severe environmental degradation can be curtailed by implications and alterations in management practices. Several mines nowadays have opted for management practices during and post-mining rehabilitation operations for closure. However, this remains limited to public safety references and degraded soils and the mine refuse are taken care of by colonized vegetation (Salazar et al. 2009; Abreu et al. 2012; Pérez-López et al. 2014; Santos et al. 2014c, 2016b; Batista et al. 2017), promoting bioattenuation of contaminants. As most of the available land remains nutrient deficient, friendly environment for plant growth is difficult (Tordoff et al. 2000; Wong 2003; Larney and Angers 2012; Santos et al. 2014a). On the contrary, intense surface drought, waterlogging, compaction, limit plant, and microbial growth (Tordoff et al. 2000), thus bioattenuation is time-consuming to achieve effective revamping of these sites (Santos et al. 2014c, 2016a).

Many soil repair techniques have been tried but most of them are either time dependent or are specific to a particular environmental setting, thus sustainable approach is required to recuperate soil (Khan et al. 2004).

3.2.8.2 Soil Rehabilitation via Ecotechnologies

Amendments Degraded soil can be managed for improvement using various organic and inorganic modifiers; however, the soil quality and its efficiency in vegetation is also dependent on other factors. Modifying materials such as sludge, sewage, and biochar can be an option but the potential risk of environmental pollution remains a concern as it contains high bacterial load, xenobiotics, and some potentially hazardous elements (PHEs) (Mendez et al. 2012; FAO and ITPS 2015). Amendments such as liming materials, biochar, hydrophilic polymers, organic matters, and metal hydroxides are common in use; these materials not only initiate element immobilization, adsorption, and/or precipitation but also supply the land with organic matter nutrients and soil minerals (Kumpiene et al. 2008). However, certain modifiers immobilize plant nutrients or enhance hazardous element uptake. To recover substrate soil cover, the use of plant refuse, livestock manure (aged or composted), etc., is in practice for ages which helps improving soil quality and supplies essential nutrients to some extent. This improves the overall quality of the soil (Larney and Angers 2012; FAO and ITPS 2015). As soil conditions improve, porosity is managed, and plant growth and germination conditions are managed. The microflora of the soil also establishes due to organic matter increase (Luna et al. 2016; Tordoff et al. 2000; Alvarenga et al. 2009; Salazar et al. 2009; Burgos et al. 2010; de Varennes et al. 2010; Larney and Angers 2012; Madeira et al. 2012). Sugar beet lime, dolomite, and quicklime, are some liming modifiers applicable in soil with acidic situation (Tordoff et al. 2000; Tiwary 2001; Wong 2003; Adriano et al. 2004; Alvarenga et al. 2008; Kumpiene et al. 2008; Burgos et al. 2010) as they enhance the pH and retard PHEs access due to precipitation as carbonates or oxyhydrides. However, they do not suffice for anion stability (Bech et al. 2014). Numerous hydrophilic polymers, having a varying composition, have been tried amendments for upgrading the water retention capability of the substrate (Clemente et al. 2004). These contain cations and complex cationic PHEs, which can enhance top layer and proper germination of plants (Qu and de Varennes 2009; de Varennes et al. 2010), but in the case of shrubs it was different (*Lavandula pedunculata* (Mill and *Cistus ladanifer* L.) Cav.) and trees (*Pistacia lentiscus* L. and *Olea europaea* var. *sylvestris* L.) (Clemente et al. 2004; Santos et al. 2013). This also upgrades the soil enzyme activity of contaminated soils (Qu and de Varennes 2009). Other similar substances such as charcoal, black carbon, biochar, and activated carbon resulting from thermal conversions can be used as an object of carbon sequestration (Ahmad et al. 2014).

Biochar, being a pocket-friendly alternative, is commonly used for soil recovery and alteration including water as it can be gathered from multiple sources including agricultural, forest waste residue, and dead animals (Méndez et al. 2012; Arán et al. 2016a; Lu et al. 2017). Being highly potent in soil recovery contaminated by PHEs, both mobilization and immobilization processes are possible in context to the element at source (Bolan et al. 2014). Immobilization in biochar increases pH, surface area, ion exchange, precipitation in the form of carbonates, oxyhydroxides, or phosphates in the soil (Cao and Harris 2010; Uchimiya et al. 2011; Puga et al. 2016). Alternatively, some oxides of metal negligible in size, including Mn, Fe, or Al, can be used to deactivate PHEs (e.g., Pb, Co, and As) in contaminated soils (Sun and Doner 1998; Sauve et al. 2000; Hartley et al. 2004; Hua et al. 2012; Madeira et al. 2012). As, goethite, and lepidocrocite, which are iron oxides/hydroxides, have double efficiency compared to clay and feldspar (Lin and Puls 2003); however, in the case of As-contaminated soils, food chain accumulation of As is lessened by goethite application (Hartley and Lepp 2008; Madeira et al. 2012). At a range of pH 4–6, the oxides are positively charged (Gimenez et al. 2007); however, in case of As (III) and As (V), adsorption pH remains a constraint.

The large surface area and the amphoteric nature of the pH-dependent sites are the key variants of nanoamendments with anion and cations (Janssen et al. 2003; Hua et al. 2012). Thus, the key factors responsible for immobilization are element and oxide (Bolan et al. 2014).

Mulching Harsh climatic conditions retard the process of mine site reclamation in arid and semi-arid areas. Mulching can be an option in these areas where wood chips, leaf shreds, or plant branches can be mounded as a fine shield to preserve soil precipitation, by limiting erosion and evaporation (Sengupta 1993; Wong 2003; Luna et al. 2016). As these mulches are rich in C and P, they tend to support germination conditions of specific plant species contributing to a superior vegetative cover (Luna et al. 2016).

Technosols Technosols are one of the fortuitous items produced globally wherever there is human intervention. By utilization of other engineering approaches, soils derived from wastes are utilized in mine site recovery. These soils are termed to be Technosols (IUSS Working Group WRB 2015). This new derived soil which can be applied in mine restoration sites must have similar patterns as to the degraded soil. This application aims at improving the degradation levels in terms of the chemical, physical, and biological qualities of the soil.

Their role in soil upgradation and maintaining a green cover has been proved by microcosm studies in laboratory (Santos et al. 2014a,b, 2016a, 2017; Rodríguez-Vila et al. 2016) and has been reported by several scientists in field-based activities (Monterroso et al. 1998; Fandiño et al. 2010; Macías et al. 2011; Asensio et al. 2013a, b, c; Bolaños et al. 2015; Arán et al. 2016b).

3.2.9 Fly Ash Application

Mining and excavation operations result in unprecedented landscape deformity and cause certain environmental and health hazards. Soil quality alterations are effects of metallic waste exposure in and around mine sites and directly impact flora, fauna, and human life. Mine spoils remain a waste as they are nutritionally deficient and are otherwise impractical for any use, unless properly treated and reclaimed.

Along with metal contaminants, the generation of acid wastes having low pH continues to raise the metal pollution levels of the soil due to increased metal solubility (Bolan et al. 2003; Lindsay et al. 2015).

Acidic leachate also termed as the AMD is an outcome of the oxidizing effects of sulfur in pyrite mine sites. This causes serious concerns for high metallic pollution in the ground and surface waters (Bolan et al. 2003; Lindsay et al. 2015; Wijesekara et al. 2016a).

Getting rid of harmful contaminants and reclamation of soils require efficient practice including chemical, biological, and engineering techniques, for achieving soil balance (Abad-Valle et al. 2017). Many reclamation strategies have been suggested by scientists, which may help in soil revamping. Table 3.6 shows different strategies used and their possible benefits.

Table 3.6 Different reclamation strategies benefits and drawbacks.

Techniques	Benefits	Drawbacks	Reference
Storing mine tailings in an isolated environment using impoundments underwater or behind dams	Avoiding exposure of mine tailings to the atmosphere	May cause a high risk to the surrounding soil and water sources	Kossoff et al. (2014), Wijesekara et al. (2016b)
Application of a cap or liner over the toxic materials	Vegetation establishment using heavy metal tolerant plant species	Breakage or tearing of the liner applied by plant roots	Lamb et al. (2014), Wijesekara et al. (2016a)
Phytostabilization	Revegetation by plant species	None	Cele and Maboeta (2016)
Inorganic alkaline materials and some clay minerals	Help in immobilizing cationic trace elements, enhance the exchange capacity of soils, reducing the solubility of trace element cations by forming strong coordination complexes with them	Performance is solely soil composition and stability dependent	Abad-Valle et al. (2017).
Biowastes application in the form of biosolids	Acts as a soil ameliorant; improves physical, chemical, and biological properties of degraded lands; enhances carbon content, soil nutrients, and microbiota.	Adequate ecotoxicological characterization must be done before application to avoid metal contaminants from sewage	Groth et al. (2016)
Fly ash (FA)	FA is used as a soil additive to improve degraded soil quality, also a source of plant micro- and macronutrients.	Metal pollutants need attention	Ahmaruzzaman (2010)

Fly ash is generated as a "by-product" from thermal power plants on burning excessive coal to fund the energy needs globally. Huge quantities of fly ash containing many metallic wastes and other toxic materials are a matter of concern due to environmental issues of pollution (Roychowdhury et al. 2016, 2018). However, this huge amount of unused fly ash can be utilized for the treatment of abandoned mine sites. Being an appropriate option for disposal beneficially, huge quantities of this ash can be utilized for a purpose, saving land requirement for dumping ash or creating ash ponds. Reclamation of mine sites using fly ash is also cost effective, as it saves funds in applying other conventional techniques. During advanced studies by researchers, application of FA as a soil additive in improvising soil quality has been acclaimed and its potential to enhance the productivity of damaged soil has also been reported. Thus it might be used as a tapped resource for soil amendment and reduce environmental risks of pollution through open dumping (Ram et al. 2006, 2007, 2008; Ahmaruzzaman 2010; Srivastava and Ram 2010; Sen and Kumar 2012; Ukwattage et al. 2013; Ram and Masto 2014).

The physiological and chemical properties of FA rely on the parent coal quality, erosion, size, age, burning techniques, and temperature at which volatilization of mineral elements takes place (Pandey and Singh 2010; Ram and Masto 2014). FA application poses a serious risk of metal contamination to soil, flora, and water bodies (surface and underground) due to chances of leaching (Carlson and Adriano 1993; Rubenstein and Segal 1993; Kukier et al. 1994).

Field studies have been initiated to estimate bulk utilization of FA in mining areas, as a large-scale application of it is required as a backfilling material in deserted mines in India. Moreover, to develop a phytocover, FA utilization is essential in former mining areas. But transportation and handling of huge amounts of FA and off-site environmental effects remain a concern. FA also contains different essential micro- and macronutrients (Ahmaruzzaman 2010) along with micro-sized particles, lower bulk density, and high fluid retention capacity, which makes it a promising candidate for being used as a soil additive. For the abandoned mine sites, FA application can be beneficial in various ways. Opencast and underground mine voids can be laid using FA in comparison to sand (Reynolds and Petrik 2005; Pradhan and Deshmukh 2008).

Being alkaline in nature, fly ash serves as a good soil layer or soil substitute acting as a liming agent. Coal fly ash enhances soil quality when ash is used in combination with soil having a pH range of 6.5–8.0 which is most demanding for growth medium. This pH is the ideal one for a growth medium. Moreover, FA is used as ash placements for recontouring and backfilling of pits. Thus, the multiutility of FA is reported in the form of capping agent, placement material to soil ameliorant (Roy et al. 2017). Agronomic property enhancing capabilities of FA has been documented by Pandey and Singh (2010) along with being an alternate material compared to lime for reclaiming acidic mine wastes (Abbott et al. 2001; Pandey and Singh 2010).

Low metal availability is an added benefit received from soil microbes along with augmented enzyme activity and N and P cycling, when soils are treated with FA sludge mixture (Jala and Goyal 2006). Increased microbial activity in FA-amended soils having added sewage sludge is also reported (Pandey and Singh 2010). However, repressive effects of toxic FA contents on soil microbiota can be mitigated by using organic mixture. Although adverse effects of FA application on microbiota are unpredictable, when FA is added in constrained amounts with organic materials, the results were affirmative, compared to higher doses of FA applications which can reduce effectiveness (Ram and Masto 2014). Acidic reactions caused due to sewage sludge can be reduced using FA in place of lime as it has a high pH value when used for ecological cleanup at mining sites. Also, an application of lignite FA upgraded the pH of the mine spoil (Ram et al. 2006; Ram and Masto 2014).

Inspite of the fact that FA is not directly related to soil microbial reform, however, when applied to the soil, it improves soil conditions which are thus amiable for microbial diversity. This also improves N and P cycling and certain enzyme actions (Ram and Masto 2014). Components of FA makes it feasible to alter sewage conditions which can be applied to sandy soils (Weber et al. 2015) as FA sewage combination enhances water stability and imparts better soil structure. Besides this technique, FA and sewage sludge mixtures can be highly productive as a soil upgrader due to their physicochemical properties and can address the management issues of both sewage sludge and FA in an eco-friendly way. Thus, biosolid wastes along with FA can enhance soil quality promoting microbial growth, reconstruct degraded soil, and amplify agricultural productivity in land constraint areas along with the utilization of degraded sites into aesthetic green covers (Gomez-Sagasti et al. 2018; Pinto et al. 2018).

3.2.10 Nanoparticle Application

Reclamation of degraded mine sites becomes very important and remains a debated issue in populated nations (Rajindiran et al. 2015). Mine refuses contain mine leachates which are critically hazardous to health and the surroundings, due to rising metal concentrations. Many processes have been introduced in this regard to remove or degrade pollutants, which include but are not limited to photochemical oxidation, coagulants, ion exchange resins, and bioremediation of mine sites for heavy metal removal. In countries such as the United States, commercialization of nanotechnological approaches to improve such techniques have been adopted (Mueller and Nowack 2010).

Nanoremediation approach utilizes nanomaterial for immobilizing and detoxification of pollutants as its fundamental, which can be either on-site or pump and treat method of remediation. Iron oxide particles (IOPs) are essential in refining biomolecules, as steel, catalysts, and pigment colors in industries and as sorbents and oxidants in environmental chemistry. The basic structures are different as they are crystalline and have different spatial arrangements. Iron oxides convert to oxides but are themselves colored in nature, and thus it imparts variations in the coloration of soils. Not only that, the presence of Fe oxides is also responsible for the surface ion adsorption, aggregation, and redox potential of the soils (Cornell and Schwertmann 2003).

3.2.10.1 Composition and Modeling of Iron Oxide Particles (IOPs)

IOPs involve various methods of synthesis such as combined precipitation, microemulsions, and hot water based reactions (Huber 2005; Laurent et al. 2008; Ali et al. 2016). Simultaneously, sol–gel conglomerate, temperature-based degradation of the original organic component, ultrasound-based reactions, hydrolysis and thermolysis of parent product, microwave technology, etc., are also followed (Mody et al. 2010). A superparamagnetic IOPs complex is produced as a result of IOPs colloidal nature, which undergoes size exclusion chromatography, magnetic filtration, and ultracentrifugation to attain a desirable size of the monodispersed particles at industrial levels. Alternately, using iron salt coprecipitation is the most convenient mode of preparing IOPs (Mascolo et al. 2013). In this process, ferric hydroxide undergoes oxidation or is aged in an aqueous medium. Generally, this takes place in the presence of ferrous salt, some alkaline components, and nitrate ions, which results in the generation of spherical magnetite IOPs having a diameter anywhere between 30 and 100 nm. The key to this recipe is the oxidation of ferrous ion and voluntary dehydration of iron hydroxides in the reaction mixture which are responsible for the precipitation of magnetite particles. The following equations summarize it:

$$FeCl_2 + NO_3^- \rightarrow Fe^{3+} + NO + H_2O\ldots \tag{3.2}$$

$$2Fe^{3+} + Fe^{2+} + 8OH^- \rightarrow Fe_3O_4 + 4H_2O\ldots \tag{3.3}$$

The above process must be carried out in oxygen-deprived conditions to avoid the formation of maghemite through magnetite oxidation (Khan et al. 2015). On one hand, the fabrication of IOPs mitigates challenges of aggregation, deprived magnetization, and biocompatibility (Mahdavi et al. 2013; Zheng et al. 2015). Additionally, functional groups are created on the nanoparticle layer which can be explored for metal adsorption. The fabrication process is smoother during coprecipitation for including chitosans which produces chitosan-coated magnetic nanoparticles. Alternately, numerous capabilities can be included in IOPs by mixing fabrication materials in later stages.

3.2.10.2 IOPs in Mine Site Reclamation

Due to efficient heavy metal adsorption properties, IOPs are quite demanding for metal removal in mine leachates. Ferrihydrite is a well-known metal-binding adsorbent due to extended surface area and unsaturated site exterior. Oxides in this category are common in existence and are found to be good absorbers of different anions and cations in soil. This type of ferrihydrite incorporation was found promising in the removal of Pb, Cd, and Cu as described by Martínez and McBride (1999). As the surface of the fabricated iron oxide particles has metal-binding functional groups, they can be effective in metal removal through adsorption techniques. Aggregation of nanoparticles is enhanced through surface alterations which initiate surface area management for smooth adsorption. In the case of magnetite nanoparticles fabricated with the amine, up to 98% of Cu from potable water was eliminated as documented (Elkady et al. 2016) and ascorbic acid coated Fe_3O_4 for effective As removal from contaminated water was reported by Feng et al. (2012). Moreover, Zhu et al. (2013) reported magnetite functionalized with silica and graphene has shown progressive adsorption toward As (Zhu et al. 2013). Alternately, red mud being a Fe_2O_3 containing composite has been useful in metal remediation at the mine site (Hizal et al. 2013) which is obtained by Bayer's process during the industry-grade process of alumina production from bauxite. Compared to others, this material promotes extensive surface reactivity and unalterable metal binding. This ensures metal-absorbed red mud as a utility element for brick making and helps in safe disposal. Additionally, it restricts metal mobility in soil and prevents leaching from mine areas. Moreover, the addition of red mud or IOPs layered with siderite prompts certain IOPs-based remediation products capable of heavy metal absorption specifically lead (Pb) (Erdem and Özverdi 2005).

3.2.10.3 Sustainable Mine Site Remediation with IOPs

Biomaterials induced with IOPs are common in practice due to their biosorption quality and are largely utilized for mine leachate retrieval using various filtration techniques, thus such sorbents which are plant derivatives such as agronomical waste derivatives, and plant or living biomasses are commercialized for mine rehabilitation due to pocket-friendly and scalability approach (Hasan and Ranjan 2010; Deepika et al. 2013; Shakoor et al. 2016).

Easy separation of biomaterials from the medium is possible as a result of their magnetization due to IOPs. Different physical and chemical strategies are adopted to induce iron oxides in biomaterials, which include electrodeposition and ball mining (Ling and Hyeon 2013; Abdul-Raheim et al. 2016). Ferrous-aluminosilicate material which is commonly available as fly ash is made of glass spheres that contain Si, Al, Fe, Na, and K primarily in its matrix. On the other hand, being caustic, fly ash is efficient in acid mine treatment. Further, induction of fly ash and IOPs complement metal adsorption qualities (Li et al. 2009; Barakat 2011).

To accomplish the treatment, fly ash undergoes an acidic soaking in 1.0 M HCl solution for 24 hours, followed by rinsing with ion-free water and subsequent drying. Further soaking in ~0.25 M ferric ion solution with a pH adjustment of 9 with the addition of NaOH is done. Later filtration of the iron-overlayed fly ash particles is processed followed by washing with ion-free water and final drying of the product at 100° C. A similar procedure of coating IOPs is applied in the case of coal bottom ash. Customized biochar coated with iron oxides is effective in metal removal from soil and water, due to induced electrostatic attraction which helps removal of As by effective binding (Inyang et al. 2016). This infusion with oxides of iron regulates the metal remediation ability of biochar, applicable largely at mine sites. The common methods of direct infusion in biochar include direct hydrolyzes, an evaporative method, and the wet impregnation method (Tan et al. 2016). However, induction of magnetic IOPs allows effortless biochar separation from the pollution treatment plant.

3.2.11 Geotechnologies

Reclamation analyzes certain measures that are taken to remodel unproductive or degraded land areas into productive and useful ones. This involves soil and land reclamation up to the root penetration levels for enhanced vegetation. However, it is quite unusual that if the underground soil seems stable, reclamation is not required. Disturbance inland structure is profoundly struck by surface mining activity which surpasses any other damage. The landforms, hydrological parameters, and many ecosystems are affected due to mining activities. Thus, reclamation of these degraded areas requires considerable attention at the grass root levels. Geotechnological reforms include soil, the materials, structures, and processes that recontour shape and surfaces of reclaimed mine lands (Favas et al. 2018).

Geotechnical considerations in surface mine reclamation influence soil and rocks that are transformed due to mining by breaking, loosening and further exposing them to chemicals and acid attacks (Prasad et al. 2018). Rock properties include intact and in situ categories that depict rock mineralogy and rock mass. While in situ properties manage rock mass patterns and alter symmetries during rock mining, intact properties get weakened by disruption. Primarily, surface mining rehabilitation depends on shear strength, compressive strength, and durability. Mining generates soil and rock structures such as pit walls, waste piles, and tailing refuse which require removal or stabilization for proper reclamation work which is usually quick and low cost. This results in steep pit walls and inactive pile slopes. Thus, planning regarding structure and erosion is required for reclamation of mine structures (Kuhn 1999).

Piling of wastes is a part and parcel of mining activities whether permanent or short termed. Excavation increases the volume of rocks which is otherwise compacted in situ. Thus, utilizing these rocks for backfilling would still result in leftovers. This material is then dumped without side slope design or compaction. In the case of mineral recoveries from ore or coal, liquid waste is generated which constitutes tailing slurry which is discharged in sludge ponds for evaporation and finally turning into supersaturated liquids, gels, and solid and requires removal which is a major challenge (Prasad et al. 2018).

Thus stabilization or management of sludge should be done based on the availability of site resources. Weathering is a natural phenomenon that cannot be averted and comprises chemical and mechanical processes which are induced by mining activities due to exposure to subsurface soils. At mining sites, AMD occurs, which is also a weathering agent.

Erosion is a well-known phenomenon that remains a matter of concern for revegetation and its impact on surface water quality, the stability of landforms, and contamination. Retrospecting the above parameters, geotechnical engineering involves design for slopes, soil covers, riprap, and other measures to resist the hassle developed during design runoff events (Kuhn 1999). Thus, the design should typically be based on worst-case assumptions; e.g. no vegetation and no infiltration of rainfall as mentioned by Kuhn (1999).

Earth fills are important aspects from the geotechnological perspective. Earth mining work, involving digging and rock and/or soil placement, is a part of site construction and mine development. Using engineering perspectives, dam and ridge constructions are done as containment for liquid wastes, cut and fill for road construction, etc. However, if the Earth fill has been designed abiding by the engineering principles and constructed with quality control, it remains sustainably stable. But unfortunately, most of the Earth fills on mine sites do not meet the designed standards of care. They were created hastily without thinking about the long-term performance. Thus, it is quite usual to find different objects such as scrap wood and metal buried in these Earth fills. Thus, for proper mine reclamation, geotechnological techniques are quite essential, and without structured strategies, mine site restoration is critical and challenging (Prasad et al. 2018).

3.3 Insights and Lessons Learned for the Practice of Mine Land Restoration

Serious environmental threats arise due to mining activities which cause pollution globally. Not only it causes soil, groundwater, and air pollution, but dangerous effects are also common on natural flora and fauna, leading to the destruction of the aesthetic view of the landscape. Rehabilitation of abandoned mine lands (AMLs) is very challenging, does require planned strategies for developing a vegetative cover on the site, and also makes the land functional for further use (Roy et al. 2017). For proper restoration, the soil quality should be compared to the original one taking care of parameters including microbial activity, plant–fungi associations, and soil enzyme activities. To compensate for the degradation, the restoration process must begin after specified mine closure activities to ensure maximum value add. Revegetation objectives should be stated mentioning post-closure land use pattern ascertained for the specific site. Ultimately, setting the self-reliant system might lead to the successful rehabilitation of AMLs (Zhenqi et al. 2012).

Mine closure activity should be highly regulated. These include mine closure planning and mine site rehabilitation including environment impact assessments which are integral parts of mine operation plans globally. To fulfil these criteria, mining organizations should set aside regulations for proper mine termination and mine site revamp. Mine closure protocol pinpoints:

- Post-mining utilization of the mining area
- Environmental renewal concept
- Scientific advancements
- Rehabilitation planning and post-mining budget
- Land return and community development

To maintain a proper rehabilitation process, issues such as soil material balance, landscape structuring, design, surface waters, and catchment areas of pre-mining status should be considered. Focus should be laid on management of agricultural and vegetation cover, panoramic-organization, wetlands management, pavement infrastructure, etc., should be well managed (Zhenqi et al. 2012). This includes execution planning and specifications treatment of dumps for reuse, agronomic development so as to reduce soil alterations, and afforestation, related health and safety documentation should all be taken care of.

3.4 Conclusion

Earlier, coal mining was the foremost preference for meeting rising energy demands without any consideration about the environment. This resulted in thousands of acres of land in a detrimental and devastating state with dangerous environmental hazards and acid drainage problems. Of late, certain policy-makers have taken into consideration, the pros and cons of mining implications and formulated policies to safeguard further damage to the landscape ecosystem. Most of the countries have now come up with detailed planning about pre-mining and post-mining regulations which requires mine operators to be more vigilant toward damaging the land and associated environment and the scenario of the mine site before and after mine closure. In the majority, it is an accepted fact that it is not always possible to return a pre-mining landscape to its original form. Though this is not the ultimate objective, still, the land should be brought back to a safe and sustainable environment. Mine site planning with aforethought of limiting damage to the adjacent landscape and environment is essential. Particularly, AMD and the leaching of trace elements and sulfates cause extensive destruction of living flora, fauna, and microbial community, and their impact remains severe. Water (subsurface and surface) affected with discoloration and increased pH also requires attention such as neutralizing with chemicals, or with the usage of indigenous rocks for their neutralizing properties over a long term.

Nowadays, it is essential to deliberate that reclamation of mine spoils and bulk utilization of FA are crucial with favorable environmental returns. FA has been quite a good choice for reclamation of mine sites as a backfilling material and treatment in AMD. FA has physical, chemical, and biological properties to act as an ameliorant to improve degraded soil qualities; this can help in sustainable vegetative cover on degraded lands. The use of geotechnological strategies including reclamation through different modes such as nanomaterials application, phytoremediation, phytostabilization, and microbial remediation is successful in remodeling the degraded land sites. Similarly, iron oxides nanoparticle is an upgraded agent in mine site remediation with proven benefits. Technological improvements of IOPs-based remediation improves metal ion adsorption and filtration process due to its biocompatibility. Earlier studies have documented that successful erosion control and soil quality upgradation are possible through plant establishment on mine tailings. Although a pre-mining state of the landscape, soil quality, and the pre-existing ecosystem are not retrieved, phytostabilization creates a self-reliant green cover that can maintain a competent soil–plant relationship. However, understanding and determining this process are time dependent. Maximum attention in mine site reclamation is showed toward agronomy and revegetation, but successful reclamation relies on the stability of underlying soil and rocks as the surface treatments greatly depend on them. Geotechnical considerations such as material properties, structures, and processes responsible for the stability have to be taken into consideration by the practitioners. Reclamation should include properties of rock and soil altered by mining activities along with water and wastes forming complex mixtures. Surface mining disrupts the geography of the existing landforms and creates pit walls, spoil and piles, tailings impoundments, and Earth fills. These need reconstruction and stabilization to promote efficient restoration. Geotechnical considerations are important for successful reclamation, and ignorance of these by practitioners may prove to be quite challenging leading to ineffective remodeling.

References

Abad-Valle, P., Iglesias-Jiménez, E., and Álvarez-Ayuso, E. (2017). A comparative study on the influence of different organic amendments on trace element mobility and microbial functionality of a polluted mine soil. *Journal of Environmental Management 188*: 287–296.

Abbott, D.E., Essington, M.E., Mullen, M.D., and Ammons, J.T. (2001). Fly ash and lime-stabilized biosolid mixtures in mine spoil reclamation: simulated weathering. *Journal of Environmental Quality 30* (2): 608–616.

Abdul-Raheim, A.R.M., El-Saeed Shimaa, M., Farag, R.K., and Abdel-Raouf Manar, E. (2016). Low cost biosorbents based on modified starch iron oxide nanocomposites for selective removal of some heavy metals from aqueous solutions. *Advanced Materials Letters 7* (5): 402–409.

Abou-Shanab, R.A.I., Angle, J.S., and Chaney, R.L. (2006). Bacterial inoculants affecting nickel uptake by Alyssum murale from low, moderate and high Ni soils. *Soil Biology and Biochemistry 38* (9): 2882–2889.

Abreu, M.M., Santos, E.S., Magalhães, M.C.F., and Fernandes, E. (2012). Trace elements tolerance, accumulation and translocation in *Cistus populifolius, Cistus salviifolius* and their hybrid growing in polymetallic contaminated mine areas. *Journal of Geochemical Exploration 123*: 52–60.

Adams, J.P., Adeli, A., Hsu, C.Y. et al. (2012). Characterization of Poplar ZIP family members ZIP1. 2 and ZNT1. *Plant Genet Transgenics* 3 (1): 1–13.

Adriano, D.C., Wenzel, W.W., Vangronsveld, J., and Bolan, N.S. (2004). Role of assisted natural remediation in environmental cleanup. *Geoderma 122* (2–4): 121–142.

Aerts, R. and Honnay, O. (2011). Forest restoration, biodiversity and ecosystem functioning. *BMC Ecology 11* (1): 1–10.

Aggangan, N., Cadiz, N., Llamado, A., and Raymundo, A. (2017). Jatropha Curcas for Bioenergy and Bioremediation in Mine Tailing Area in Mogpog, Marinduque, Philippines. *Energy Procedia* 110: 471–478. https://doi.org/10.1016/J.EGYPRO.2017.03.171.

Ahemad, M. and Khan, M.S. (2011). Functional aspects of plant growth promoting rhizobacteria: recent advancements. *Insight Microbiology 1* (3): 39–54.

Ahmad, M., Rajapaksha, A.U., Lim, J.E. et al. (2014). Biochar as a sorbent for contaminant management in soil and water: a review. *Chemosphere 99*: 19–33.

Ahmaruzzaman, M. (2010). A review on the utilization of fly ash. *Progress in Energy and Combustion Science 36* (3): 327–363.

Al Agely, A., Sylvia, D.M., and Ma, L.Q. (2005). Mycorrhizae increase arsenic uptake by the hyperaccumulator Chinese brake fern (Pteris vittata L.). *Journal of Environmental Quality 34* (6): 2181–2186.

Ali, H., Khan, E., and Sajad, M.A. (2013). Phytoremediation of heavy metals—concepts and applications. *Chemosphere 91* (7): 869–881.

Ali, A., Hira Zafar, M.Z., ul Haq, I. et al. (2016). Synthesis, characterization, applications, and challenges of iron oxide nanoparticles. *Nanotechnology, Science and Applications 9*: 49.

Alloway, B.J. (1995). *Heavy Metals in Soils*, 2ee. London: Blackie Academicand Professional.

Alshaal, T., Domokos-Szabolcsy, É., Márton, L. et al. (2013). Phytoremediation of bauxite-derived red mud by giant reed. *Environmental Chemistry Letters 11* (3): 295–302.

Alvarenga, P., Palma, P., Gonçalves, A.P. et al. (2008). Assessment of chemical, biochemical and ecotoxicological aspects in a mine soil amended with sludge of either urban or industrial origin. *Chemosphere 72* (11): 1774–1781.

Alvarenga, P., Palma, P., Gonçalves, A.P. et al. (2009). Organic residues as immobilizing agents in aided phytostabilization:(II) effects on soil biochemical and ecotoxicological characteristics. *Chemosphere 74* (10): 1301–1308.

Anderson, C.W., Brooks, R.R., Stewart, R.B., and Simcock, R. (1998). Harvesting a crop of gold in plants. *Nature 395* (6702): 553–554.

Anderson, C., Brooks, R., Stewart, R. et al. (1999a). The phytoremediation and phytomining of heavy metals. *Pacrim* 99: 127–135.

Anderson, C.W., Stewart, R.B., Moreno, F.N. et al. (2003). Gold phytomining. Novel developments in a plant-based mining system. In: *Proceedings of the Gold 2003 Conference: New Industrial Applications of Gold*. World Gold Council and Canadian Institute of Mining, Metallurgy and Petroleum.

Anderson, J.D., Ingram, L.J., and Stahl, P.D. (2008). Influence of reclamation management practices on microbial biomass carbon and soil organic carbon accumulation in semiarid mined lands of Wyoming. *Applied Soil Ecology 40* (2): 387–397.

Antoun, H., Beauchamp, C.J., Goussard, N. et al. (1998). Potential of Rhizobium and Bradyrhizobium species as plant growth promoting rhizobacteria on non-legumes: effect on radishes (Raphanus sativus L.). In: *Molecular Microbial Ecology of the Soil* (ed. G. Hardarson and W.J. Broughton), 57–67. Dordrecht: Springer.

Arán, D., Antelo, J., Fiol, S., and Macías, F. (2016a). Influence of feedstock on the copper removal capacity of waste-derived biochars. *Bioresource Technology 212*: 199–206.

Arán, D., Cortizas, J.M.A., and Vázquez, F.M. (2016b). Uso de Tecnosuelos para la mejora en la calidad química de aguas de escorrentía de la mina Fé (Cuidad Rodrigo, Salamanca). In: *Solos e Água, fontes (esgotáveis) de vida e de desenvolvimento Livro de atas: VII Congresso Ibérico das Ciências do Solo (CICS 2016) VI Congresso Nacional de Rega e Drenagem, 13-15 de Setembro de 2016, Instituto Politécnico de Beja, Beja*, 337–340. Sociedade Portuguesa da Ciência do Solo (SPCS).

Aranda, J.M., O'connor, G.A., and Eiceman, G.A. (1989). *Effects of Sewage Sludge on Di-(2-ethylhexyl) Phthalate Uptake by Plants* (Vol. 18, No. 1, pp. 45–50). American Society of Agronomy, Crop Science Society of America, and Soil Science Society of America.

Asensio, V., Vega, F.A., Andrade, M.L., and Covelo, E.F. (2013a). Technosols made of wastes to improve physico-chemical characteristics of a copper mine soil. *Pedosphere 23* (1): 1–9.

Asensio, V., Vega, F.A., Andrade, M.L., and Covelo, E.F. (2013b). Tree vegetation and waste amendments to improve the physical condition of copper mine soils. *Chemosphere 90* (2): 603–610.

Asensio, V., Covelo, E.F., and Kandeler, E. (2013c). Soil management of copper mine tailing soils—sludge amendment and tree vegetation could improve biological soil quality. *Science of the Total Environment 456*: 82–90.

Asmelash, F., Bekele, T., and Birhane, E. (2016). The potential role of arbuscular mycorrhizal fungi in the restoration of degraded lands. *Frontiers in Microbiology 7*: 1095.

Augé, R.M., Toler, H.D., and Saxton, A.M. (2015). Arbuscular mycorrhizal symbiosis alters stomatal conductance of host plants more under drought than under amply watered conditions: a meta-analysis. *Mycorrhiza 25* (1): 13–24.

Bąba, W., Błońska, A., Kompała-Bąba, A. et al. (2016). Arbuscular mycorrhizal fungi (AMF) root colonization dynamics of Molinia caerulea (L.) Moench. in grasslands and post-industrial sites. *Ecological Engineering 95*: 817–827.

Badri, D.V., Weir, T.L., Van der Lelie, D., and Vivanco, J.M. (2009). Rhizosphere chemical dialogues: plant–microbe interactions. *Current Opinion in Biotechnology 20* (6): 642–650.

Baker, A.J.M., McGrath, S.P., Reeves, R.D., and Smith, J.A.C. (2000). Metal hyperaccumulator plants: a review of the ecology and physiology of a biological resource for phytoremediation of metal-polluted soils. In: *Phytoremediation of Contaminated Soil and Water* (ed. N. Terry and G.S. Banuelos), 408. Boca Raton, FL: CRC Press.

Baker, A., Brooks, R., and Reeves, R. (1988). Growing for gold... and copper... and zinc. *New Scientist 10* (1603): 44–48.

Balat, M. (2007). Influence of coal as an energy source on environmental pollution. *Energy Sources, Part A: Recovery, Utilization, and Environmental Effects 29* (7): 581–589.

Bang, J., Kamala-Kannan, S., Lee, K.J. et al. (2015). Phytoremediation of heavy metals in contaminated water and soil using Miscanthus sp. Goedae-Uksae 1. *International Journal of Phytoremediation 17* (6): 515–520.

Barakat, M.A. (2011). New trends in removing heavy metals from industrial wastewater. *Arabian Journal of Chemistry 4* (4): 361–377.

Barbosa, B., Boléo, S., Sidella, S. et al. (2015). Phytoremediation of heavy metal-contaminated soils using the perennial energy crops Miscanthus spp. and Arundo donax L. *BioEnergy Research 8* (4): 1500–1511.

Barbosa, B., Costa, J., Boléo, S. et al. (2016). Phytoremediation of inorganic compounds. In: *Electrokinetics Across Disciplines and Continents – New Strategies for Sustainable Development* (ed. A.B. Ribeiro, E.P. Mateus and N. Couto), 373–400. Switzerland: *Springer International Publishing*.

Barceló, J., Vázquez, M.D., Mádico, J., and Poschenrieder, C. (1994). Hyperaccumulation of zinc and cadmium in Thlaspi caerulescens. In: *Environmental Contamination*, 132–134. Edinburgh: CEP Consultants Ltd.

Barnhisel, R.I. and Hower, J.M. (1997). Coal surface mine reclamation in the eastern United States: the revegetation of disturbed lands to hayland/pasture or cropland. *Advances in Agronomy* 61: 233–275.

Batista, M.J., Gonzalez-Fernandez, O., Abreu, M.M. et al. (2017). Pioneer mediterranean shrub species revegetating soils developed on mining soils/spoils. *Land Degradation & Development 28* (2): 718–730.

Becerra-Castro, C., Monterroso, C., Prieto-Fernández, A. et al. (2012). Pseudometallophytes colonising Pb/Zn mine tailings: a description of the plant–microorganism–rhizosphere soil system and isolation of metal-tolerant bacteria. *Journal of Hazardous Materials 217*: 350–359.

Bech, J., Poschenrieder, C., Barceló, J., and Lansac, A. (2002). Plants from mine spoils in the South American area as potential sources of germplasm for phytoremediation technologies. *Acta Biotechnologica* 22 (1–2): 5–11. 10.1002/1521-3846(200205)22:1/2<5::AID-ABIO5>3.0.CO;2-B.

Bech, J., Abreu, M.M., Chon, H.T., and Roca, N. (2014). Remediation of potentially toxic elements in contaminated soils. In: *PHEs, Environment and Human Health*, 253–308. Dordrecht: Springer.

Bellin, C.A. and O'Connor, G.A. (1990). *Plant Uptake of Pentachlorophenol from Sludge-Amended Soils* (Vol. 19, No. 3, pp. 598–602). American Society of Agronomy, Crop Science Society of America, and Soil Science Society of America.

Beneduzi, A., Ambrosini, A., and Passaglia, L.M. (2012). Plant growth-promoting rhizobacteria (PGPR): their potential as antagonists and biocontrol agents. *Genetics and Molecular Biology 35* (4): 1044–1051.

Berendsen, R.L., Pieterse, C.M., and Bakker, P.A. (2012). The rhizosphere microbiome and plant health. *Trends in Plant Science 17* (8): 478–486.

Berg, G. (2009). Plant–microbe interactions promoting plant growth and health: perspectives for controlled use of microorganisms in agriculture. *Applied Microbiology and Biotechnology 84* (1): 11–18.

Berg, G. and Smalla, K. (2009). Plant species and soil type cooperatively shape the structure and function of microbial communities in the rhizosphere. *FEMS Microbiology Ecology 68* (1): 1–13.

Berg, G., Zachow, C., Müller, H. et al. (2013). Next-generation bio-products sowing the seeds of success for sustainable agriculture. *Agronomy* 3 (4): 648–656. doi: 10.3390/agronomy3040648.

Bhalerao, S.A. (2013). Arbuscular mycorrhizal fungi: a potential biotechnological tool for phytoremediation of heavy metal contaminated soils. *International Journal of Science and Nature 4* (1): 1–15.

Bhattacharya, T., Banerjee, D.K., and Gopal, B. (2006). Heavy metal uptake by Scirpus littoralis schrad. from fly ash dosed and metal spiked soils. *Environmental Monitoring and Assessment 121* (1): 363–380.

Bhattacharyya, P.N. and Jha, D.K. (2012). Plant growth-promoting rhizobacteria (PGPR): emergence in agriculture. *World Journal of Microbiology and Biotechnology 28* (4): 1327–1350.

Bishnoi, U. (2015). PGPR interaction: an ecofriendly approach promoting the sustainable agriculture system. *Advances in Botanical Research 75*: 81–113.

Bolan, N.S., Adriano, D.C., and Curtin, D. (2003). Soil acidification and liming interactions with nutrient and heavy metal transformation and bioavailability. *Advances in Agronomy 78* (21): 5–272.

Bolan, N.S., Park, J.H., Robinson, B. et al. (2011). Phytostabilization: a green approach to contaminant containment. *Advances in Agronomy 112*: 145–204.

Bolan, N., Kunhikrishnan, A., Thangarajan, R. et al. (2014). Remediation of heavy metal (loid) s contaminated soils–to mobilize or to immobilize? *Journal of Hazardous Materials 266*: 141–166.

Bolaños, D., García, F.M., Verde, J.R., and Macías, F. (2015). Aplicación de Tecnosoles para la mitigación de efectos ambientales por construcción civil en zonas de pizarras. *Congreso de Ciencia y Tecnología ESPE* 10 (1): 117–122.

Bonfante, P. and Anca, I.A. (2009). Plants, mycorrhizal fungi, and bacteria: a network of interactions. *Annual Review of Microbiology 63*: 363–383.

Borie, F., Rubio, R., and Morales, A. (2008). Arbuscular mycorrhizal fungi and soil aggregation. *Journal of Soil Science and Plant Nutrition* 8 (2): 9–18.

Bottini, R., Cassán, F., and Piccoli, P. (2004). Gibberellin production by bacteria and its involvement in plant growth promotion and yield increase. *Applied Microbiology and Biotechnology 65* (5): 497–503.

Bouillant, M.L., Miché, L., Ouedraogo, O. et al. (1997). Inhibition of Striga seed germination associated with sorghum growth promotion by soil bacteria. *Comptes Rendus de l'Académie des Sciences-Series III-Sciences de la Vie 320* (2): 159–162.

Bradshaw, A. (1997). Restoration of mined lands—using natural processes. *Ecological Engineering 8* (4): 255–269.

Bradshaw, A. and Johnson, M. (1992). Revegetation of metalliferous mine waste: the range of practical techniques used in Western Europe. *International Conference Minerals, Metals and the Environment* 481–491.

Brazil, G.M., Kenefick, L., Callanan, M. et al. (1995). Construction of a rhizosphere pseudomonad with potential to degrade polychlorinated biphenyls and detection of bph gene expression in the rhizosphere. *Applied and Environmental Microbiology 61* (5): 1946–1952.

Broeckling, C.D., Broz, A.K., Bergelson, J. et al. (2008). Root exudates regulate soil fungal community composition and diversity. *Applied and Environmental Microbiology 74* (3): 738–744.

Brooks, R.R., Chambers, M.F., Nicks, L.J., and Robinson, B.H. (1998). Phytomining. *Trends in Plant Science 3* (9): 359–362.

Brunori, A.M.E., Sdringola, P., Dini, F. et al. (2017). Carbon balance and life cycle assessment in an oak plantation for mined area reclamation. *Journal of Cleaner Production 144*: 69–78.

Burgos, P., Madejón, P., Cabrera, F., and Madejón, E. (2010). By-products as amendment to improve biochemical properties of trace element contaminated soils: effects in time. *International Biodeterioration & Biodegradation 64* (6): 481–488.

Cabral, L., Siqueira, J.O., Soares, C.R.F.S., and Pinto, J.E.B.P. (2010). Retenção de metais pesados em micélio de fungos micorrízicos arbusculares. *Química Nova 33* (1): 25–29.

Cao, X. and Harris, W. (2010). Properties of dairy-manure-derived biochar pertinent to its potential use in remediation. *Bioresource Technology 101* (14): 5222–5228.

Carlson, C.L. and Adriano, D.C. (1993). Environmental impacts of coal combustion residues. *Journal of Environmental Quality 22* (2): 227–247.

Caterino, B., Schuler, J., Grushecky, S., and Skousen, J. (2017). Surface mine to biomass farm: growing shrub willow (salix spp.) In northeastern west virginia-first year results. *Journal American Society of Mining and Reclamation 6* (1).

Cattaneo, F., Di Gennaro, P., Barbanti, L. et al. (2014). Perennial energy cropping systems affect soil enzyme activities and bacterial community structure in a South European agricultural area. *Applied Soil Ecology 84*: 213–222.

Cattelan, A.J., Hartel, P.G., and Fuhrmann, J.J. (1999). Screening for plant growth–promoting rhizobacteria to promote early soybean growth. *Soil Science Society of America Journal 63* (6): 1670–1680.

Cele, E.N. and Maboeta, M. (2016). A greenhouse trial to investigate the ameliorative properties of biosolids and plants on physicochemical conditions of iron ore tailings: Implications for an iron ore mine site remediation. *Journal of Environmental Management 165*: 167–174.

Chabot, R., Beauchamp, C.J., Kloepper, J.W., and Antoun, H. (1998). Effect of phosphorus on root colonization and growth promotion of maize by bioluminescent mutants of phosphate-solubilizing Rhizobium leguminosarum biovar phaseoli. *Soil Biology and Biochemistry 30* (12): 1615–1618.

Chamber of Mines of South Africa (2008). Facts and figures 2008.

Chandler, D., Davidson, G., Grant, W.P. et al. (2008). Microbial biopesticides for integrated crop management: an assessment of environmental and regulatory sustainability. *Trends in Food Science & Technology 19* (5): 275–283.

Chang, A.C., Page, A.L., and Koo, B.J. (2002). Biogeochemistry of phosphorus, iron, and trace elements in soils as influenced by soil-plant-microbial interactions. *Developments in Soil Science* 28: 43–57.

Chen, B.D., Zhu, Y.G., and Smith, F.A. (2006). Effects of arbuscular mycorrhizal inoculation on uranium and arsenic accumulation by Chinese brake fern (Pteris vittata L.) from a uranium mining-impacted soil. *Chemosphere 62* (9): 1464–1473.

Cheung, K.C., Wong, J.P.K., Zhang, Z.Q. et al. (2000). Revegetation of lagoon ash using the legume species Acacia auriculiformis and Leucaena leucocephala. *Environmental Pollution 109* (1): 75–82.

Chodak, M., Pietrzykowski, M., and Sroka, K. (2015). Physiological profiles of microbial communities in mine soils afforested with different tree species. *Ecological Engineering 81*: 462–470.

Clemente, A.S., Werner, C., Máguas, C. et al. (2004). Restoration of a limestone quarry: effect of soil amendments on the establishment of native Mediterranean sclerophyllous shrubs. *Restoration Ecology 12* (1): 20–28.

Collins, G.B. and Petolino, J.F. (2012). *Applications of Genetic Engineering to Crop Improvement*, vol. 10. Springer Science & Business Media.

Conesa, H.M., García, G., Faz, Á., and Arnaldos, R. (2007). Dynamics of metal tolerant plant communities' development in mine tailings from the Cartagena-La Unión Mining District (SE Spain) and their interest for further revegetation purposes. *Chemosphere 68* (6): 1180–1185.

Cornell, R.M. and Schwertmann, U. (2003). *The Iron Oxides: Structure, Properties, Reactions, Occurrences and Uses*. John Wiley & Sons.

Costa, J., Barbosa, B., and Fernando, A.L. (2016). Wastewaters reuse for energy crops cultivation. In: *Technological Innovation for Cyber-Physical Systems* (ed. L.M. Camarinha-Matos, A.J. Falcão, N. Vafaei and S. Najdi). 7th IFIP WG 5.5/ SOCOLNET Advanced Doctoral Conference on Computing, Electrical and Industrial Systems, DoCEIS 2016, Costa de Caparica, Portugal, April 11–13, 2016, Proceedings, pp. 507–514, Springer, Cham.

Cunningham, S.D., Berti, W.R., and Huang, J.W. (1995). Phytoremediation of contaminated soils. *Trends in Biotechnology 13* (9): 393–397.

Dadon, T., Nun, N.B., and Mayer, A.M. (2004). A factor from Azospirillum brasilense inhibits germination and radicle growth of Orobanche aegyptiaca. *Israel Journal of Plant Sciences 52* (2): 83–86.

Danesh, Y.R., Tajbakhsh, M., Goltapeh, E.M., and Varma, A. (2013). Mycoremediation of heavy metals. In: *Fungi as Bioremediators* (ed. E.M. Goltapeh, Y.R. Danesh and A. Varma), 245–267. Berlin, Heidelberg: Springer.

Deepika, C., Dipak, S., and Anjani, P. (2013). Heavy metal remediation of wastewater by agro wastes–a review. *European Chemical Bulletin 2* (11): 880–886.

Dell'Amico, E., Cavalca, L., and Andreoni, V. (2005). Analysis of rhizobacterial communities in perennial Graminaceae from polluted water meadow soil, and screening of metal-resistant, potentially plant growth-promoting bacteria. *FEMS Microbiology Ecology 52* (2): 153–162.

Dhal, B., Thatoi, H.N., Das, N.N., and Pandey, B.D. (2013). Chemical and microbial remediation of hexavalent chromium from contaminated soil and mining/ metallurgical solid waste: a review. *Journal of Hazardous Materials 250*: 272–291.

Diop, T.A. (2003). in vitro culture of arbuscular mycorrhizal fungi: advances and future prospects. *African Journal of Biotechnology 2* (12): 692–697.

Djurdjević, L., Mitrović, M., Pavlović, P. et al. (2006). Phenolic acids as bioindicators of fly ash deposit revegetation. *Archives of Environmental Contamination and Toxicology 50* (4): 488–495.

Dobbelaere, S., Croonenborghs, A., Thys, A. et al. (1999). Phytostimulatory effect of Azospirillum brasilense wild type and mutant strains altered in IAA production on wheat. *Plant and Soil 212* (2): 153–162.

Dodd, I.C., Zinovkina, N.Y., Safronova, V.I., and Belimov, A.A. (2010). Rhizobacterial mediation of plant hormone status. *Annals of Applied Biology 157* (3): 361–379.

Dongmei, L.E.I. and Changqun, D.U.A.N. (2008). Restoration potential of pioneer plants growing on lead-zinc mine tailings in Lanping, southwest China. *Journal of Environmental Sciences 20* (10): 1202–1209.

Dorner, J. (2002). An introduction to using native plants in restoration projects. Center for Urban Horticulture, University of Washington, Seattle. http://www. nps. gov/plants/restore/pubs/intronatplant (accessed 12-09-2021).

Doty, S.L. (2008). Enhancing phytoremediation through the use of transgenics and endophytes. *New Phytologist 179* (2): 318–333.

Dowarah, J., Boruah, H.D., Gogoi, J. et al. (2009). Eco-restoration of a high-sulphur coal mine overburden dumping site in northeast India: a case study. *Journal of Earth System Science 118* (5): 597–608.

Dutta, R.K. and Agrawal, M. (2002). Effect of tree plantations on the soil characteristics and microbial activity of coal mine spoil land. *Tropical Ecology 43* (2): 315–324.

Dželetović, Ž.S. and Filipović, R. (1995). Grain characteristics of crops grown on power plant ash and bottom slag deposit. *Resources, Conservation and Recycling 13* (2): 105–113.

Dželetović, Ž.S., Filipović, R.M., Stojanović, D.D., and Lazarević, M.M. (2009). Impact of lignite washery sludge on mine soil quality and poplar trees growth. *Land Degradation & Development 20* (2): 145–155.

Elkady, M., Hassan, H.S., and Hashim, A. (2016). Immobilization of magnetic nanoparticles onto amine-modified nano-silica gel for copper ions remediation. *Materials 9* (6): 460.

Environmental Law Alliance Worldwide (ELAW) (2010). *Guidebook for Evaluating Mining Project EIAs*. Eugene, Oregon: *Environmental Law Alliance Worldwide*.

Erakhrumen, A.A. and Agbontalor, A. (2007). Phytoremediation: an environmentally sound technology for pollution prevention, control and remediation in developing countries. *Educational Research and Review 2* (7): 151–156.

Erdem, M. and Özverdi, A. (2005). Lead adsorption from aqueous solution onto siderite. *Separation and Purification Technology 42* (3): 259–264.

Ernst, W.H. (2005). Phytoextraction of mine wastes–options and impossibilities. *Geochemistry 65*: 29–42.

Facchetti, S., Balasso, A., Fichtner, C. et al. (1986). Studies on the absorption of TCDD by some plants species. *Chemosphere (Oxford) 15* (9–12): 1387–1388.

Fan, D., Liu, T., Li, C. et al. (2015). Efficient CRISPR/Cas9-mediated targeted mutagenesis in Populus in the first generation. *Scientific Reports 5* (1): 1–7.

Fandiño, V.A., Couce, M.A., Vega, F.A., and Covelo, E.F. (2010). Characterization of different age technosols developed on a copper mine tailing. *Fresenius Environmental Bulletin 19* (8b): 1687–1693.

FAO and ITPS (2015). Status of the world's soil resources (SWSR)–main report. *Food and Agriculture Organization of the United Nations and Intergovernmental Technical Panel on Soils*, Rome, Italy, 650. ISBN 978-92-5-109004-6.

Favas, P.J., Martino, L.E., and Prasad, M.N. (2018). Abandoned mine land reclamation—challenges and opportunities (Holistic Approach). In: *Bio-Geotechnologies for Mine Site Rehabilitation*, 3–31. Elsevier.

Fellet, G., Marchiol, L., Delle Vedove, G., and Peressotti, A. (2011). Application of biochar on mine tailings: effects and perspectives for land reclamation. *Chemosphere 83* (9): 1262–1267.

Feng, L., Cao, M., Ma, X. et al. (2012). Superparamagnetic high-surface-area Fe_3O_4 nanoparticles as adsorbents for arsenic removal. *Journal of Hazardous Materials 217*: 439–446.

Fernando, A.L., Duarte, M.P., Almeida, J. et al. (2010). Environmental impact assessment of energy crops cultivation in Europe. *Biofuels, Bioproducts and Biorefining 4* (6): 594–604.

Fernando, A.L., Costa, J., Barbosa, B. et al. (2018). Environmental impact assessment of perennial crops cultivation on marginal soils in the Mediterranean Region. *Biomass and Bioenergy 111*: 174–186.

Finlay, R.D. (2008). Ecological aspects of mycorrhizal symbiosis: with special emphasis on the functional diversity of interactions involving the extraradical mycelium. *Journal of Experimental Botany 59* (5): 1115–1126.

Firmin, S., Labidi, S., Fontaine, J. et al. (2015). Arbuscular mycorrhizal fungal inoculation protects Miscanthus× giganteus against trace element toxicity in a highly metal-contaminated site. *Science of the Total Environment 527*: 91–99.

Fladung, M. and Polak, O. (2012). Ac/Ds-transposon activation tagging in poplar: a powerful tool for gene discovery. *BMC Genomics 13* (1): 1–16.

Flores-Tavizón, E., Alarcón-Herrera, M.T., González-Elizondo, S., and Olguin, E.J. (2003). Arsenic tolerating plants from mine sites and hot springs in the semi-arid region of Chihuahua, Mexico. *Acta Biotechnologica 23* (2–3): 113–119.

Förstner, U. and Wittmann, G.T. (2012). *Metal Pollution in the Aquatic Environment*. Springer Science & Business Media.

Freitas, E.V., Nascimento, C.W., Souza, A., and Silva, F.B. (2013). Citric acid-assisted phytoextraction of lead: a field experiment. *Chemosphere 92* (2): 213–217.

Fuentes-Ramirez, L.E. and Caballero-Mellado, J. (2005). Bacterial biofertilizers. In: *PGPR: Biocontrol and Biofertilization* (ed. Z.A. Siddiqui), 143–172. Dordrecht, The Netherlands: Springer.

Gajić, G., Djurdjević, L., Kostić, O. et al. (2016). Assessment of the phytoremediation potential and an adaptive response of Festuca rubra L. sown on fly ash deposits: Native grass has a pivotal role in ecorestoration management. *Ecological Engineering 93*: 250–261.

Gajić, G., Djurdjević, L., Kostić, O. et al. (2018). Ecological potential of plants for phytoremediation and ecorestoration of fly ash deposits and mine wastes. *Frontiers in Environmental Science 6*: 124.

Gao, L., Miao, Z., Bai, Z. et al. (1998). A case study of ecological restoration at the Xiaoyi Bauxite Mine, Shanxi Province, China. *Ecological Engineering 11* (1–4): 221–229.

Gaur, A. and Adholeya, A. (2004). Prospects of arbuscular mycorrhizal fungi in phytoremediation of heavy metal contaminated soils. *Current Science 86*: 528–534.

Gaur, A. and Varma, A. (2007). Research methods in arbuscular mycorrhizal fungi. In: *Advanced Techniques in Soil Microbiology* (ed. A. Varma and R. Oelmüller), 377–396. Berlin, Heidelberg: Springer.

Germaine, K.J., Liu, X., Cabellos, G.G. et al. (2006). Bacterial endophyte-enhanced phytoremediation of the organochlorine herbicide 2, 4-dichlorophenoxyacetic acid. *FEMS Microbiology Ecology 57* (2): 302–310.

Ghnaya, T., Mnassri, M., Ghabriche, R. et al. (2015). Nodulation by Sinorhizobium meliloti originated from a mining soil alleviates Cd toxicity and increases Cd-phytoextraction in Medicago sativa L. *Frontiers in Plant Science 6*: 863.

Giménez, J., Martínez, M., de Pablo, J. et al. (2007). Arsenic sorption onto natural hematite, magnetite, and goethite. *Journal of Hazardous Materials 141* (3): 575–580.

Glick, B.R. (2012). Plant growth-promoting bacteria: mechanisms and applications. *Scientifica 2012*.

Glick, B.R., Penrose, D.M., and Li, J. (1998). A model for the lowering of plant ethylene concentrations by plant growth-promoting bacteria. *Journal of Theoretical Biology 190* (1): 63–68.

Gómez-Sagasti, M.T., Hernández, A., Artetxe, U. et al. (2018). How valuable are organic amendments as tools for the phytomanagement of degraded soils? The knowns, known unknowns, and unknowns. *Frontiers in Sustainable Food Systems 2*: 68. doi: 10.3389/fsufs.2018.00068.

González, Á., García-Gonzalo, P., Gil-Díaz, M.M. et al. (2019). Compost-assisted phytoremediation of As-polluted soil. *Journal of Soils and Sediments 19* (7): 2971–2983.

Gonzalez-Chavez, C., D'haen, J., Vangronsveld, J., and Dodd, J.C. (2002a). Copper sorption and accumulation by the extraradical mycelium of different Glomus spp. (arbuscular mycorrhizal fungi) isolated from the same polluted soil. *Plant and Soil 240* (2): 287–297.

Gonzalez-Chavez, C., Harris, P.J., Dodd, J., and Meharg, A.A. (2002b). Arbuscular mycorrhizal fungi confer enhanced arsenate resistance on Holcus lanatus. *New Phytologist 155* (1): 163–171.

González-Oreja, J.A., Rozas, M.A., Alkorta, I., and Garbisu, C. (2008). Dendroremediation of heavy metal polluted soils. *Reviews on Environmental Health 23* (3): 223.

Goswami, D., Thakker, J.N., and Dhandhukia, P.C. (2016). Portraying mechanics of plant growth promoting rhizobacteria (PGPR): a review. *Cogent Food & Agriculture 2* (1): 1127500.

Grandlic, C.J., Mendez, M.O., Chorover, J. et al. (2008). Plant growth-promoting bacteria for phytostabilization of mine tailings. *Environmental Science & Technology 42* (6): 2079–2084.

Gray, E.J. and Smith, D.L. (2005). Intracellular and extracellular PGPR: commonalities and distinctions in the plant–bacterium signaling processes. *Soil Biology and Biochemistry 37* (3): 395–412.

Grime, J.P. (1979). *Plant Strategies and Vegetation Processes Vegetation Processes*. John Wiley & Sons, Limited.

Grobelak, A. (2016). Organic soil amendments in the phytoremediation process. In: *Phytoremediation* (ed. A. Ansari, S. Gill, R. Gill, et al.), 21–39. Cham: Springer 10.1007/978-3-319-41811-7_2.

Groninger, J., Skousen, J., Angel, P. et al. (2007). Mine reclamation practices to enhance forest development through natural succession. *The Appalachian Regional Reforestation Initiative (ARRI). Forest Reclamation Advisory No. 5* 1–5.

Groth, V.A., Carvalho-Pereira, T., da Silva, E.M., and Niemeyer, J.C. (2016). Ecotoxicological assessment of biosolids by microcosms. *Chemosphere 161*: 342–348.

Grotz, N. and Guerinot, M.L. (2006). Molecular aspects of Cu, Fe and Zn homeostasis in plants. *Biochimica et Biophysica Acta (BBA)-Molecular. Cell Research 1763* (7): 595–608.

Grünewald, H., Böhm, C., Quinkenstein, A. et al. (2009). Robinia pseudoacacia L.: a lesser known tree species for biomass production. *BioEnergy Research 2* (3): 123–133.

Guo, Z.H. and Miao, X.F. (2010). Growth changes and tissues anatomical characteristics of giant reed (Arundo donax L.) in soil contaminated with arsenic, cadmium and lead. *Journal of Central South University of Technology 17* (4): 770–777.

Haferburg, G. and Kothe, E. (2010). Metallomics: lessons for metalliferous soil remediation. *Applied Microbiology and Biotechnology 87* (4): 1271–1280.

Hartley, W. and Lepp, N.W. (2008). Remediation of arsenic contaminated soils by iron-oxide application, evaluated in terms of plant productivity, arsenic and phytotoxic metal uptake. *Science of the Total Environment 390* (1): 35–44.

Hartley, W., Edwards, R., and Lepp, N.W. (2004). Arsenic and heavy metal mobility in iron oxide-amended contaminated soils as evaluated by short-and long-term leaching tests. *Environmental Pollution 131* (3): 495–504.

Hasan, S.H. and Ranjan, D. (2010). Agro-industrial waste: a low-cost option for the biosorptive remediation of selenium anions. *Industrial & Engineering Chemistry Research 49* (19): 8927–8934.

Hassinen, V.H., Tervahauta, A.I., Schat, H., and Kärenlampi, S.O. (2011). Plant metallothioneins–metal chelators with ROS scavenging activity? *Plant Biology 13* (2): 225–232.

Haynes, R.J. (2009). Reclamation and revegetation of fly ash disposal sites – challenges and research needs. *Journal of Environmental Management 90* (1): 43–53.

He, J., Ma, C., Ma, Y. et al. (2013). Cadmium tolerance in six poplar species. *Environmental Science and Pollution Research 20* (1): 163–174.

He, J., Li, H., Ma, C. et al. (2015). Overexpression of bacterial γ-glutamylcysteine synthetase mediates changes in cadmium influx, allocation and detoxification in poplar. *New Phytologist 205* (1): 240–254.

Herman, D.J., Firestone, M.K., Nuccio, E., and Hodge, A. (2012). Interactions between an arbuscular mycorrhizal fungus and a soil microbial community mediating litter decomposition. *FEMS Microbiology Ecology 80* (1): 236–247.

Hettiarachchi, G.M., Pierzynski, G.M., and Ransom, M.D. (2001). In situ stabilization of soil lead using phosphorus. *Journal of Environmental Quality 30* (4): 1214–1221.

Hizal, J., Tutem, E., Guclu, K. et al. (2013). Heavy metal removal from water by red mud and coal fly ash: an integrated adsorption–solidification/stabilization process. *Desalination and Water Treatment 51* (37–39): 7181–7193.

Hodge, A. and Fitter, A.H. (2010). Substantial nitrogen acquisition by arbuscular mycorrhizal fungi from organic material has implications for N cycling. *Proceedings of the National Academy of Sciences 107* (31): 13754–13759.

Holl, K.D. (2002). Long-term vegetation recovery on reclaimed coal surface mines in the eastern USA. *Journal of Applied Ecology 39* (6): 960–970.

Hua, M., Zhang, S., Pan, B. et al. (2012). Heavy metal removal from water/wastewater by nanosized metal oxides: a review. *Journal of Hazardous Materials 211*: 317–331.

Huang, X.D., El-Alawi, Y., Gurska, J. et al. (2005). A multi-process phytoremediation system for decontamination of persistent total petroleum hydrocarbons (TPHs) from soils. *Microchemical Journal 81* (1): 139–147.

Huang, X.F., Chaparro, J.M., Reardon, K.F. et al. (2014). Rhizosphere interactions: root exudates, microbes, and microbial communities. *Botany 92* (4): 267–275.

Huber, D.L. (2005). Synthesis, properties, and applications of iron nanoparticles. *Small 1* (5): 482–501.

Inyang, M.I., Gao, B., Yao, Y. et al. (2016). A review of biochar as a low-cost adsorbent for aqueous heavy metal removal. *Critical Reviews in Environmental Science and Technology 46* (4): 406–433.

IUSS Working Group WRB (2015). World reference base for soil resources 2014, update 2015 International soil classification system for naming soils and creating legends for soil maps. *World Soil Resources Report, 106*. FAO, Rome.

Ivanova, L.A., Ronzhina, D.A., Ivanov, L.A. et al. (2011). Over-expression of gsh1 in the cytosol affects the photosynthetic apparatus and improves the performance of transgenic poplars on heavy metal-contaminated soil. *Plant Biology 13* (4): 649–659.

Jala, S. and Goyal, D. (2006). Fly ash as a soil ameliorant for improving crop production—a review. *Bioresource Technology 97* (9): 1136–1147.

Jan, S., Rashid, B., Azooz, M.M. et al. (2016). Genetic strategies for advancing phytoremediation potential in plants: a recent update. *Plant Metal Interaction* 431–454.

Janssen, R.P.T., Bruggenwert, M.G.M., and Van Riemsdijk, W.H. (2003). Zinc ion adsorption on montmorillonite–Al hydroxide polymer systems. *European Journal of Soil Science 54* (2): 347–355.

Jasper, D.A. (2007). Beneficial soil microorganisms of the jarrah forest and their recovery in bauxite mine restoration in Southwestern Australia. *Restoration Ecology 15*: S74–S84.

Jeffries, P., Gianinazzi, S., Perotto, S. et al. (2003). The contribution of arbuscular mycorrhizal fungi in sustainable maintenance of plant health and soil fertility. *Biology and Fertility of Soils 37* (1): 1–16.

Jiang, C.Y., Sheng, X.F., Qian, M., and Wang, Q.Y. (2008). Isolation and characterization of a heavy metal-resistant Burkholderia sp. from heavy metal-contaminated paddy field soil and its potential in promoting plant growth and heavy metal accumulation in metal-polluted soil. *Chemosphere 72* (2): 157–164.

Joner, E.J., Briones, R., and Leyval, C. (2000). Metal-binding capacity of arbuscular mycorrhizal mycelium. *Plant and Soil 226* (2): 227–234.

Kanzler, M., Böhm, C., and Freese, D. (2015). Impact of P fertilisation on the growth performance of black locust (Robinia pseudoacacia L.) in a lignite post-mining area in Germany. *Annals of. Forest Research 58* (1): 39–54.

Karbowska, B. (2016). Presence of thallium in the environment: sources of contaminations, distribution and monitoring methods. *Environmental Monitoring and Assessment 188* (11): 1–19.

Kavourides, K., Pavloudakis, F. and Filios, F. (2002). Environmental protection and land reclamation works in West Macedonia Lignite Centre–Current practice and future perspectives. *SWEMP 2002: Proceedings of the 7th International Symposium on Environmental Issues and Waste Management in Energy and Mineral Production*, Cagliari (October). pp. 627–634.

Khan, A.G. (2005). Role of soil microbes in the rhizospheres of plants growing on trace metal contaminated soils in phytoremediation. *Journal of Trace Elements in Medicine and Biology 18* (4): 355–364.

Khan, A.G., Kuek, C., Chaudhry, T.M. et al. (2000). Role of plants, mycorrhizae and phytochelators in heavy metal contaminated land remediation. *Chemosphere 41* (1-2): 197–207.

Khan, F.I., Husain, T., and Hejazi, R. (2004). An overview and analysis of site remediation technologies. *Journal of Environmental Management 71* (2): 95–122.

Khan, U.S., Manan, A., Khan, N. et al. (2015). Transformation mechanism of magnetite nanoparticles. *Materials Science-Poland 33* (2): 278–285.

Kim, J.G. and Dixon, J.B. (2002). Oxidation and fate of chromium in soils. *Soil Science and Plant Nutrition 48* (4): 483–490.

Kleczewski, N.M., Herms, D.A., and Bonello, P. (2012). Nutrient and water availability alter belowground patterns of biomass allocation, carbon partitioning, and ectomycorrhizal abundance in Betula nigra. *Trees 26* (2): 525–533.

Kloepper, J.W. and Schroth M.N. (1978). Plant growth-promoting rhizobacteria on radishes. In: Angers, A.B. and Clarey, G., (Ed.) *Proceedings of the 4th International Conference on Plant Pathogenic Bacteria*, Tennesse, New York, pp. 879–882.

Kloepper, J.W., Ryu, C.M., and Zhang, S. (2004). Induced systemic resistance and promotion of plant growth by Bacillus spp. *Phytopathology 94* (11): 1259–1266.

Koprivova, A., Kopriva, S., Jäger, D. et al. (2002). Evaluation of transgenic poplars over-expressing enzymes of glutathione synthesis for phytoremediation of cadmium. *Plant Biology 4* (6): 664–670.

Kossoff, D., Dubbin, W.E., Alfredsson, M. et al. (2014). Mine tailings dams: characteristics, failure, environmental impacts, and remediation. *Applied Geochemistry 51*: 229–245.

Kost, D.A., Vimmerstedt, J.P., and Brown, J.H. (1998). Site factor effects on establishment of planted and volunteer trees and shrubs on graded cast overburden and replaced topsoil in southeastern Ohio. *International Journal of Surface Mining, Reclamation and Environment 12* (2): 79–86.

Kostić, O., Mitrović, M., Knežević, M. et al. (2012). The potential of four woody species for the revegetation of fly ash deposits from the'Nikola Tesla-a'thermoelectric plant (Obrenovac, Serbia). *Archives of Biological Sciences 64* (1): 145–158.

Kramer, P.A., Zabowski, D., Scherer, G., and Everett, R.L. (2000). Native plant restoration of copper mine tailings: II. *Field Survival, Growth, and Nutrient Uptake* (Vol. 29, No. 6, pp. 1770–1777). American Society of Agronomy, Crop Science Society of America, and Soil Science Society of America.

Krzaklewski, W., Pietrzykowski, M., and Woś, B. (2012). Survival and growth of alders (Alnus glutinosa (L.) Gaertn. and Alnus incana (L.) Moench) on fly ash technosols at different substrate improvement. *Ecological Engineering 49*: 35–40.

Kuhn, A. (1999). Geotechnical considerations in surface mine reclamation. In: *Proceedings 16th Annual National Meeting of the American Society for Surface Mining Reclamation "Mining and Reclamation in the next Millennium"* (ed. S.A. Bengson and D.M. Bland), 13–19. Scottsdale, Arizona: ASSMR.

Kukier, U., Sumner, M.E., and Miller, W.P. (1994). *Boron Release from Fly Ash and its Uptake by Corn* (Vol. 23, No. 3, pp. 596–603). American Society of Agronomy, Crop Science Society of America, and Soil Science Society of America.

Kumar, A. and Maiti, S.K. (2015). Effect of organic manures on the growth of Cymbopogon citratus and Chrysopogon zizanioides for the phytoremediation of chromite-asbestos mine waste: a pot scale experiment. *International Journal of Phytoremediation* 17: 437–447. https://doi.org/10.1080/15226514.2014.910174.

Kumar, A., Usmani, Z., and Kumar, V. (2017). Biochar and flyash inoculated with plant growth promoting rhizobacteria act as potential biofertilizer for luxuriant growth and yield of tomato plant. *Journal of Environmental Management 190*: 20–27.

Kumpiene, J., Lagerkvist, A., and Maurice, C. (2008). Stabilization of As, Cr, Cu, Pb and Zn in soil using amendments–a review. *Waste Management 28* (1): 215–225.

Lamb, D.T., Venkatraman, K., Bolan, N. et al. (2014). Phytocapping: an alternative technology for the sustainable management of landfill sites. *Critical Reviews in Environmental Science and Technology 44* (6): 561–637.

Larney, F.J. and Angers, D.A. (2012). The role of organic amendments in soil reclamation: a review. *Canadian Journal of Soil Science 92* (1): 19–38.

Larondelle, N. and Haase, D. (2012). Valuing post-mining landscapes using an ecosystem services approach—an example from Germany. *Ecological Indicators 18*: 567–574.

Laurent, S., Forge, D., Port, M. et al. (2008). Magnetic iron oxide nanoparticles: synthesis, stabilization, vectorization, physicochemical characterizations, and biological applications. *Chemical Reviews 108* (6): 2064–2110.

Lee, S.H., Ji, W., Lee, W.S. et al. (2014). Influence of amendments and aided phytostabilization on metal availability and mobility in Pb/Zn mine tailings. *Journal of Environmental Management 139*: 15–21.

Lei, H., Peng, Z., Yigang, H., and Yang, Z. (2016). Vegetation and soil restoration in refuse dumps from open pit coal mines. *Ecological Engineering 94*: 638–646.

Leung, H.M., Wu, F.Y., Cheung, K.C. et al. (2010a). Synergistic effects of arbuscular mycorrhizal fungi and phosphate rock on heavy metal uptake and accumulation by an arsenic hyperaccumulator. *Journal of Hazardous Materials 181* (1–3): 497–507.

Leung, H.M., Wu, F.Y., Cheung, K.C. et al. (2010b). The effect of arbuscular mycorrhizal fungi and phosphate amendement on arsenic uptake, accumulation and growth of *Pteris vittata* in As-contaminated soil. *International Journal of Phytoremediation 12* (4): 384–403.

Leyval, C., Joner, E.J., Del Val, C., and Haselwandter, K. (2002). Potential of arbuscular mycorrhizal fungi for bioremediation. In: *Mycorrhizal Technology in Agriculture* (ed. S. Gianinazzi, H. Schüepp, J.M. Barea and K. Haselwandter), 175–186. Basel: Birkhäuser 10.1007/978-3-0348-8117-3_14.

Li, M.S. (2006). Ecological restoration of mineland with particular reference to the metalliferous mine wasteland in China: a review of research and practice. *Science of the Total Environment 357* (1–3): 38–53.

Li, X. and Huang, L. (2015). Toward a new paradigm for tailings phytostabilization—nature of the substrates, amendment options, and anthropogenic pedogenesis. *Critical Reviews in Environmental Science and Technology 45* (8): 813–839.

Li, J. and Zhang, Y. (2012). Remediation technology for the uranium contaminated environment: a review. *Procedia Environmental Sciences 13*: 1609–1615.

Li, L., Fan, M., Brown, R.C. et al. (2009). Production of a new wastewater treatment coagulant from fly ash with concomitant flue gas scrubbing. *Journal of Hazardous Materials 162* (2-3): 1430–1437.

Li, C., Wang, X., Ran, L. et al. (2015). PtoMYB92 is a transcriptional activator of the lignin biosynthetic pathway during secondary cell wall formation in Populus tomentosa. *Plant and Cell Physiology 56* (12): 2436–2446.

Lima, A.T., Mitchell, K., O'Connell, D.W. et al. (2016). The legacy of surface mining: remediation, restoration, reclamation and rehabilitation. *Environmental Science & Policy 66*: 227–233.

Lin, Z. and Puls, R.W. (2003). Potential indicators for the assessment of arsenic natural attenuation in the subsurface. *Advances in Environmental Research 7* (4): 825–834.

Lindsay, M.B., Moncur, M.C., Bain, J.G. et al. (2015). Geochemical and mineralogical aspects of sulfide mine tailings. *Applied Geochemistry 57*: 157–177.

Ling, D. and Hyeon, T. (2013). Chemical design of biocompatible iron oxide nanoparticles for medical applications. *Small 9* (9–10): 1450–1466.

Liu, D., Jiang, W., Liu, C. et al. (2000). Uptake and accumulation of lead by roots, hypocotyls and shoots of Indian mustard [Brassica juncea (L.)]. *Bioresource Technology 71* (3): 273–277.

Liu, J., Li, S., Ouyang, Z. et al. (2008). Ecological and socioeconomic effects of China's policies for ecosystem services. *Proceedings of the National Academy of Sciences 105* (28): 9477–9482.

Lottermoser, B.G. (2010). *Mine Wastes: Characterization, Treatment And Environmental Impacts*. Springer, Berlin, Germany: ISBN 978-3-642-12418-1, 10.1007/978-3-642-12419-8.

Lu, K., Yang, X., Gielen, G. et al. (2017). Effect of bamboo and rice straw biochars on the mobility and redistribution of heavy metals (Cd, Cu, Pb and Zn) in contaminated soil. *Journal of Environmental Management 186*: 285–292.

Lubbers, I.M., Van Groenigen, K.J., Brussaard, L., and Van Groenigen, J.W. (2015). Reduced greenhouse gas mitigation potential of no-tillage soils through earthworm activity. *Scientific Reports 5* (1): 1–10.

Lugtenberg, B. and Kamilova, F. (2009). Plant-growth-promoting rhizobacteria. *Annual Review of Microbiology 63*: 541–556.

Lugtenberg, B.J., Chin-A-Woeng, T.F., and Bloemberg, G.V. (2002). Microbe–plant interactions: principles and mechanisms. *Antonie Van Leeuwenhoek 81* (1): 373–383.

Luna, L., Pastorelli, R., Bastida, F. et al. (2016). The combination of quarry restoration strategies in semiarid climate induces different responses in biochemical and microbiological soil properties. *Applied Soil Ecology 107*: 33–47.

Luo, J., Shi, W., Li, H. et al. (2016). The conserved salt-responsive genes in the roots of *Populus× canescens* and *Arabidopsis thaliana. Environmental and Experimental Botany* 129: 48–56.

Ma, Y., Prasad, M.N.V., Rajkumar, M., and Freitas, H. (2011). Plant growth promoting rhizobacteria and endophytes accelerate phytoremediation of metalliferous soils. *Biotechnology Advances 29* (2): 248–258.

Ma, Y., He, J., Ma, C. et al. (2014). Ectomycorrhizas with P axillus involutus enhance cadmium uptake and tolerance in P opulus× canescens. *Plant, Cell & Environment 37* (3): 627–642.

Ma, Y., Rajkumar, M., Zhang, C., and Freitas, H. (2016). Beneficial role of bacterial endophytes in heavy metal phytoremediation. *Journal of Environmental Management 174*: 14–25.

Macías, F., Macías-García, F., Nieto, C. et al. (2011). Gestión de residuos y cambio climático. In: *Gestión de residuos orgánicos de uso agrícola* (ed. M.E.L. Mosquera and M.J.S. Osés), 11–24. Santiago de Compostela: Servizo de Publicacións e Intercambio Científico de la Universidade de Santiago de Compostela.

Madeira, A.C., De Varennes, A., Abreu, M.M. et al. (2012). Tomato and parsley growth, arsenic uptake and translocation in a contaminated amended soil. *Journal of Geochemical Exploration 123*: 114–121.

Maestri, E. and Marmiroli, N. (2011). Transgenic plants for phytoremediation. *International Journal of Phytoremediation 13* (supl): 264–279.

Mahdavi, M., Ahmad, M.B., Haron, M.J. et al. (2013). Synthesis, surface modification and characterisation of biocompatible magnetic iron oxide nanoparticles for biomedical applications. *Molecules 18* (7): 7533–7548.

Mains, D., Craw, D., Rufaut, C.G., and Smith, C.M.S. (2006). Phytostabilization of gold mine tailings, New Zealand. Part 1: plant establishment in alkaline saline substrate. *International Journal of Phytoremediation 8* (2): 131–147.

Maiti, S.K. (2007). Bioreclamation of coalmine overburden dumps—with special empasis on micronutrients and heavy metals accumulation in tree species. *Environmental Monitoring and Assessment 125* (1): 111–122.

Maiti, S.K. (2012). *Ecorestoration of the Coalmine Degraded Lands*. Springer Science & Business Media.

Maiti, S.K. (2013). Ecology and ecosystem in mine-degraded land. In: *Ecorestoration of the Coalmine Degraded Lands*, 21–37. India: Springer 10.1007/978-81-322-0851-8_2.

Maiti, S.K. and Jaiswal, S. (2008). Bioaccumulation and translocation of metals in the natural vegetation growing on fly ash lagoons: a field study from Santaldih thermal power plant, West Bengal, India. *Environmental Monitoring and Assessment* 136: 355–370. https://doi.org/10.1007/s10661-007-9691-5.

Maksimovic, I., Kastori, R., Putnik-Delic, M., and Boršev, M. (2014). Effect of yttrium on photosynthesis and water relations in young maize plants. *Journal of Rare Earths 32* (4): 372–378.

Malayeri, B.E., Chehregani, A., Mohsenzadeh, F. et al. (2013). Plants growing in a mining area: screening for metal accumulator plants possibly useful for bioremediation. *Toxicological and Environmental Chemistry* 95: 434–444. doi: 10.1080/02772248.2013.788701.

Malik, A. (2004). Metal bioremediation through growing cells. *Environment International 30* (2): 261–278.

Mardis, E.R. (2017). DNA sequencing technologies: 2006–2016. *Nature Protocols 12* (2): 213.

Marmiroli, M., Pietrini, F., Maestri, E. et al. (2011). Growth, physiological and molecular traits in Salicaceae trees investigated for phytoremediation of heavy metals and organics. *Tree Physiology 31* (12): 1319–1334.

Marques, A.P., Oliveira, R.S., Rangel, A.O., and Castro, P.M. (2006). Zinc accumulation in Solanum nigrum is enhanced by different arbuscular mycorrhizal fungi. *Chemosphere 65* (7): 1256–1263.

Martínez, C.E. and McBride, M.B. (1999). Dissolved and labile concentrations of Cd, Cu, Pb, and Zn in aged ferrihydrite– organic matter systems. *Environmental Science & Technology 33* (5): 745–750.

Mascolo, M.C., Pei, Y., and Ring, T.A. (2013). Room temperature co-precipitation synthesis of magnetite nanoparticles in a large pH window with different bases. *Materials 6* (12): 5549–5567.

Meier, S., Borie, F., Bolan, N., and Cornejo, P. (2012a). Phytoremediation of metal-polluted soils by arbuscular mycorrhizal fungi. *Critical Reviews in Environmental Science and Technology 42* (7): 741–775.

Mench, M., Lepp, N., Bert, V. et al. (2010). Successes and limitations of phytotechnologies at field scale: outcomes, assessment and outlook from COST Action 859. *Journal of Soils and Sediments 10* (6): 1039–1070.

Mendes, R., Garbeva, P., and Raaijmakers, J.M. (2013). The rhizosphere microbiome: significance of plant beneficial, plant pathogenic, and human pathogenic microorganisms. *FEMS Microbiology Reviews 37* (5): 634–663.

Mendez, M.O. and Maier, R.M. (2008). Phytoremediation of mine tailings in temperate and arid environments. *Reviews in Environmental Science and Bio/Technology 7* (1): 47–59.

Méndez, A., Gómez, A., Paz-Ferreiro, J., and Gascó, G. (2012). Effects of sewage sludge biochar on plant metal availability after application to a Mediterranean soil. *Chemosphere 89* (11): 1354–1359.

Mendoza, G.A. (2000). *GIS-based Multicriteria Approaches to Land Use Suitability Assessment and Allocation*, 89–94. United States Department of Agriculture Forest Service General Technical Report NC.

Mendez, M.O., Glenn, E.P., and Maier, R.M. (2007). Phytostabilization potential of quailbush for mine tailings: growth, metal accumulation, and microbial community changes. *Journal of Environmental Quality* 36: 245–253.

Miao, Y.A.N.G., Xiao, X.Y., Miao, X.F. et al. (2012). Effect of amendments on growth and metal uptake of giant reed (Arundo donax L.) grown on soil contaminated by arsenic, cadmium and lead. *Transactions of Nonferrous Metals Society of China 22* (6): 1462–1469.

Migeon, A., Blaudez, D., Wilkins, O. et al. (2010). Genome-wide analysis of plant metal transporters, with an emphasis on poplar. *Cellular and Molecular Life Sciences 67* (22): 3763–3784.

Mills, R.F., Krijger, G.C., Baccarini, P.J. et al. (2003). Functional expression of AtHMA4, a P1B-type ATPase of the Zn/Co/Cd/Pb subclass. *The Plant Journal 35* (2): 164–176.

Miransari, M. (2010). Contribution of arbuscular mycorrhizal symbiosis to plant growth under different types of soil stress. *Plant Biology 12* (4): 563–569.

Miransari, M. (2011). Interactions between arbuscular mycorrhizal fungi and soil bacteria. *Applied Microbiology and Biotechnology 89* (4): 917–930.

Mirza, N., Pervez, A., Mahmood, Q. et al. (2011). Ecological restoration of arsenic contaminated soil by Arundo donax L. *Ecological Engineering 37* (12): 1949–1956.

Misra, S. and Gedamu, L. (1989). Heavy metal tolerant transgenic Brassica napus L. and Nicotiana tabacum L. plants. *Theoretical and Applied Genetics 78* (2): 161–168.

Mitrović, M., Pavlović, P., Lakušić, D. et al. (2008). The potential of Festuca rubra and Calamagrostis epigejos for the revegetation of fly ash deposits. *Science of the Total Environment 407* (1): 338–347.

Mody, V.V., Siwale, R., Singh, A., and Mody, H.R. (2010). Introduction to metallic nanoparticles. *Journal of Pharmacy and Bioallied Sciences 2* (4): 282.

Monterroso, C., Macias, F., Gil Bueno, A., and Val Caballero, C. (1998). Evaluation of the land reclamation project at the As Pontes Mine (NW Spain) in relation to the suitability of the soil for plant growth. *Land Degradation & Development 9* (5): 441–451.

Moore, F.P., Barac, T., Borremans, B. et al. (2006). Endophytic bacterial diversity in poplar trees growing on a BTEX-contaminated site: the characterisation of isolates with potential to enhance phytoremediation. *Systematic and Applied Microbiology 29* (7): 539–556.

Morandi, F., Perrin, A., and Østergård, H. (2016). Miscanthus as energy crop: environmental assessment of a miscanthus biomass production case study in France. *Journal of Cleaner Production 137*: 313–321.

Moreno, F.N., Anderson, C.W., Stewart, R.B., and Robinson, B.H. (2008). Phytofiltration of mercury-contaminated water: volatilisation and plant-accumulation aspects. *Environmental and Experimental Botany 62* (1): 78–85.

Mpofu, K., Mushiri, S., and Mushiri, K. (2013). An investigation into the effectiveness of coal ash in acid mine drainage (AMD) abatement. A case study of iron duke mine. *International Researcher* 2 (3): 49–65.

Mueller, N.C. and Nowack, B. (2010). Nanoparticles for remediation: solving big problems with little particles. *Elements 6* (6): 395–400.

Mukhopadhyay, S. and Maiti, S.K. (2011). Trace metal accumulation and natural mycorrhizal colonisation in an afforested coalmine overburden dump: a case study from India. *International Journal of Mining, Reclamation and Environment 25* (2): 187–207.

Mukhopadhyay, S., Maiti, S.K., and Masto, R.E. (2013). Use of Reclaimed Mine Soil Index (RMSI) for screening of tree species for reclamation of coal mine degraded land. *Ecological Engineering 57*: 133–142.

Mulligan, C.N., Yong, R.N., and Gibbs, B.F. (2001). Remediation technologies for metal-contaminated soils and groundwater: an evaluation. *Engineering Geology 60* (1–4): 193–207.

Mummey, D.L., Stahl, P.D., and Buyer, J.S. (2002). Microbial biomarkers as an indicator of ecosystem recovery following surface mine reclamation. *Applied Soil Ecology 21* (3): 251–259.

Munshower, F.F. (1994). *Practical Handbook of Disturbed Land Revegetation*. Boca Raton, FL: *Lewis Publishing*.

Nadeem, S.M., Ahmad, M., Zahir, Z.A. et al. (2014). The role of mycorrhizae and plant growth promoting rhizobacteria (PGPR) in improving crop productivity under stressful environments. *Biotechnology Advances 32* (2): 429–448.

Negri, M.C., Hinchman, R.R., and Gatliff, E.G. (1996). *Phytoremediation: Using Green Plants to Clean up Contaminate Soil, Groundwater, and Wastewater (No. ANL/ES/CP-89941; CONF-960804-38)*. IL, USA: Argonne National Lab.

Neri, A.C. and Sánchez, L.E. (2010). A procedure to evaluate environmental rehabilitation in limestone quarries. *Journal of Environmental Management 91* (11): 2225–2237.

Neuschütz, C. and Greger, M. (2009). Ability of various plant species to prevent leakage of N, P, and metals from sewage sludge. *International Journal of Phytoremediation 12* (1): 67–84.

Nirola, R., Megharaj, M., Beecham, S. et al. (2016). Remediation of metalliferous mines, revegetation challenges and emerging prospects in semi-arid and arid conditions. *Environmental Science and Pollution Research 23* (20): 20131–20150.

Nowack, B., Schulin, R., and Luster, J. (2010). Metal fractionation in a contaminated soil after reforestation: temporal changes versus spatial variability. *Environmental Pollution 158* (10): 3272–3278.

NRC (National Research Council) (2005). *Mineral Tolerance of Animals*, 2 revisede. Washington, DC: National Academies Press.

Nriagu, J.O. (1996). A history of global metal pollution. *Science 272* (5259): 223–223.

Nsanganwimana, F., Pourrut, B., Mench, M., and Douay, F. (2014). Suitability of Miscanthus species for managing inorganic and organic contaminated land and restoring ecosystem services. A review. *Journal of Environmental Management 143*: 123–134.

Nurtjahya, E., Setiadi, D., Guhardja, E., and Setiadi, Y. (2008). Establishment of four native tree species for potential revegetating of Tin-Mined Land in Bangka Island, Indonesia. In: *Proceedings of the Third International Seminar on Mine Closure*, 751–758. Australian Centre for Geomechanics.

Orłowska, E., Godzik, B., and Turnau, K. (2012). Effect of different arbuscular mycorrhizal fungal isolates on growth and arsenic accumulation in Plantago lanceolata L. *Environmental Pollution 168*: 121–130.

Palmer, M.A., Bernhardt, E.S., Schlesinger, W.H. et al. (2010). Mountaintop mining consequences. *Science 327* (5962): 148–149.

Pandey, V.C. and Singh, N. (2010). Impact of fly ash incorporation in soil systems. *Agriculture, Ecosystems & Environment 136* (1–2): 16–27.

Pandey, V.C., Pandey, D.N., and Singh, N. (2015a). Sustainable phytoremediation based on naturally colonizing and economically valuable plants. *Journal of Cleaner Production 86*: 37–39.

Pandey, V.C., Prakash, P., Bajpai, O. et al. (2015b). Phytodiversity on fly ash deposits: evaluation of naturally colonized species for sustainable phytorestoration. *Environmental Science and Pollution Research 22* (4): 2776–2787.

Pandey, V.C., Bajpai, O., and Singh, N. (2016a). Plant regeneration potential in fly ash ecosystem. *Urban Forestry & Urban Greening 15*: 40–44.

Papazoglou, E.G. (2007). Arundo donax L. stress tolerance under irrigation with heavy metal aqueous solutions. *Desalination 211* (1–3): 304–313.

Pavel, P.B., Puschenreiter, M., Wenzel, W.W. et al. (2014). Aided phytostabilization using Miscanthus sinensis× giganteus on heavy metal-contaminated soils. *Science of the Total Environment 479*: 125–131.

Pavlović, P., Mitrović, M., and Djurdjević, L. (2004). An ecophysiological study of plants growing on the fly ash deposits from the "Nikola Tesla–A" thermal power station in Serbia. *Environmental Management 33* (5): 654–663.

Peng, R., Lin, G., and Li, J. (2016). Potential pitfalls of CRISPR/Cas9-mediated genome editing. *The FEBS journal 283* (7): 1218–1231.

Pérez-López, R., Márquez-García, B., Abreu, M.M. et al. (2014). Erica andevalensis and Erica australis growing in the same extreme environments: phytostabilization potential of mining areas. *Geoderma 230*: 194–203.

Perrig, D., Boiero, M.L., Masciarelli, O.A. et al. (2007). Plant-growth-promoting compounds produced by two agronomically important strains of Azospirillum brasilense, and implications for inoculant formulation. *Applied Microbiology and Biotechnology 75* (5): 1143–1150.

Pietrzykowski, M., Socha, J., and van Doorn, N.S. (2015). Scots pine (Pinus sylvestris L.) site index in relation to physico-chemical and biological properties in reclaimed mine soils. *New Forests 46* (2): 247–266.

del Pilar Ortega-Larrocea, M., Xoconostle-Cazares, B., Maldonado-Mendoza, I.E. et al. (2010). Plant and fungal biodiversity from metal mine wastes under remediation at Zimapan, Hidalgo, Mexico. *Environmental Pollution 158* (5): 1922–1931.

Pineda, A., Zheng, S.J., van Loon, J.J. et al. (2010). Helping plants to deal with insects: the role of beneficial soil-borne microbes. *Trends in Plant Science 15* (9): 507–514.

Pinto, A.P., de Varennes, A., Castanheiro, J.E., and Balsinhas, A.M. (2018). Fly ash and lime-stabilized biosolid mixtures in mine spoil reclamation. In: *Bio-geotechnologies for mine site rehabilitation* (ed. M.N.V. Prasad, P.J. de Campos Favas and S.K. Maiti), 159–180. Elsevier.

Prach, K. and Hobbs, R.J. (2008). Spontaneous succession versus technical reclamation in the restoration of disturbed sites. *Restoration Ecology 16* (3): 363–366. doi: https://doi.org/10.1111/j.1526-100X.2008.00412.x.

Pradhan, A. and Deshmukh, J.P. (2008). Utilization of fly ash for treatment of acid mine water. *Journal of Environmental Research and Development 3* (1): 137–142.

Prasad, M.N.V. (2003). Metal hyperaccumulation in plants-Biodiversity prospecting forphytoremediation technology. *Electronic Journal of Biotechnology 6*: 110–146.

Prasad, M.N.V. (2006). Trace elements in medicinal plants: environmental quality and human health concerns. *Inżynieria Ekologiczna 16*: 56–58.

Prasad, M.N.V., de Campos Favas, P.J., and Maiti, S.K. (ed.) (2018). *Bio-geotechnologies for Mine Site Rehabilitation*. Elsevier.

Puga, A.P., Melo, L.C.A., de Abreu, C.A. et al. (2016). Leaching and fractionation of heavy metals in mining soils amended with biochar. *Soil and Tillage Research 164*: 25–33.

Pusztahelyi, T., Holb, I.J., and Pócsi, I. (2015). Secondary metabolites in fungus-plant interactions. *Frontiers in Plant Science 6*: 573.

Qu, G. and de Varennes, A. (2009). Use of hydrophilic insoluble polymers in the restoration of metal-contaminated soils. *Applied and Environmental Soil Science* 2009: 1–8. doi: 10.1155/2009/790687.

Quinkenstein, A. and Jochheim, H. (2016). Assessing the carbon sequestration potential of poplar and black locust short rotation coppices on mine reclamation sites in Eastern Germany–Model development and application. *Journal of Environmental Management 168*: 53–66.

Rai, U.N., Pandey, K., Sinha, S. et al. (2004). Revegetating fly ash landfills with Prosopis juliflora L.: impact of different amendments and Rhizobium inoculation. *Environment International 30* (3): 293–300.

Rajindiran, S., Dotaniya, M.L., Coumar, M.V. et al. (2015). Heavy metal polluted soils in India: status and countermeasures. *JNKVV Research Journal 49* (3): 320–337.

Ram, L.C. and Masto, R.E. (2014). Fly ash for soil amelioration: a review on the influence of ash blending with inorganic and organic amendments. *Earth-Science Reviews 128*: 52–74.

Ram, L.C., Srivastava, N.K., Tripathi, R.C. et al. (2006). Management of mine spoil for crop productivity with lignite fly ash and biological amendments. *Journal of Environmental Management 79* (2): 173–187.

Ram, L.C., Srivastava, N.K., Jha, S.K. et al. (2007). Management of lignite fly ash for improving soil fertility and crop productivity. *Environmental Management 40* (3): 438–452.

Ram, L.C., Jha, S.K., Tripathi, R.C. et al. (2008). Remediation of fly ash landfills through plantation. *Remediation Journal: The Journal of Environmental Cleanup Costs, Technologies & Techniques 18* (4): 71–90.

Randall, B.K., Yost, M.A., Kitchen, N.R. et al. (2016). Impact of rhizome quality on Miscanthus establishment in claypan soil landscapes. *Industrial Crops and Products 85*: 331–340.

Reddy, G.N. and Prasad, M.N.V. (1990). Heavy metal-binding proteins/peptides: occurrence, structure, synthesis and functions. A review. *Environmental and Experimental Botany 30* (3): 251–264.

Redecker, D., Kodner, R., and Graham, L.E. (2000). Glomalean fungi from the Ordovician. *Science 289* (5486): 1920–1921.

Reverchon, F., Yang, H., Ho, T.Y. et al. (2015). A preliminary assessment of the potential of using an acacia—biochar system for spent mine site rehabilitation. *Environmental Science and Pollution Research 22* (3): 2138–2144.

Reynolds, K. and Petrik, L. (2005, April). The use of fly ash for the control and treatment of acid mine drainage. In: *Proceedings of the 2004 Water Institute of Southern Africa (WISA) Biennial Conference*, Cape Town, South Africa (2–6 May 2004), Organized by Event Dynamics, ISBN: 1-920-01728-3, Produced by: Document Transformation Technologies.

Richards, R.T., Chambers, J.C., and Ross, C. (1998). Use of native plants on federal lands: policy and practice. *Rangeland Ecology & Management/Journal of Range Management Archives 51* (6): 625–632.

Richardson, A.E. (2001). Prospects for using soil microorganisms to improve the acquisition of phosphorus by plants. *Functional Plant Biology 28* (9): 897–906.

Riefler, M., Novak, O., Strnad, M., and Schmülling, T. (2006). Arabidopsis cytokinin receptor mutants reveal functions in shoot growth, leaf senescence, seed size, germination, root development, and cytokinin metabolism. *The Plant Cell 18* (1): 40–54.

Rillig, M.C. (2004). Arbuscular mycorrhizae and terrestrial ecosystem processes. *Ecology Letters 7* (8): 740–754.

Rimmer, L.D. and Younger, A. (1997). Land reclamation after coal-mining operations. In: *Contaminated Land and its Reclamation* (ed. R.E. Hester and R.M. Harrison), 73–90. Royal Society of Chemistry.

Robinson, B.H., Brooks, R.R., and Clothier, B.E. (1999). Soil amendments affecting nickel and cobalt uptake by Berkheya coddii: potential use for phytomining and phytoremediation. *Annals of Botany 84* (6): 689–694.

Rodriguez, L., Lopez-Bellido, F.J., Carnicer, A. et al. (2005). Mercury recovery from soils by phytoremediation. In: *Environmental Chemistry*, 197–204. Berlin, Heidelberg: Springer.

Rodríguez, H., Fraga, R., Gonzalez, T., and Bashan, Y. (2006). Genetics of phosphate solubilization and its potential applications for improving plant growth-promoting bacteria. *Plant and Soil 287* (1): 15–21.

Rodríguez-Vila, A., Asensio, V., Forján, R., and Covelo, E.F. (2016). Assessing the influence of technosol and biochar amendments combined with Brassica juncea L. on the fractionation of Cu, Ni, Pb and Zn in a polluted mine soil. *Journal of Soils and Sediments 16* (2): 339–348.

Roy, M., Roychowdhury, R., and Mukherjee, P. (2017). *Phytoreclamation of Abandoned Acid Mine Drainage Site After Treatment with Fly Ash*. IntechOpen.

Roychowdhury, R., Mukherjee, P., and Roy, M. (2016). Identification of chromium resistant bacteria from dry fly ash sample of Mejia MTPS thermal power plant, West Bengal, India. *Bulletin of Environmental Contamination and Toxicology 96* (2): 210–216.

Roychowdhury, R., Roy, M., Rakshit, A. et al. (2018). Arsenic bioremediation by indigenous heavy metal resistant bacteria of fly ash pond. *Bulletin of Environmental Contamination and Toxicology 101* (4): 527–535.

Rubenstein, R. and Segal, S.A. (1993). Risk assessment of metals in groundwater. *Metals in Groundwater* 209–308.

Rufyikiri, G., Thiry, Y., Wang, L. et al. (2002). Uranium uptake and translocation by the arbuscular mycorrhizal fungus, Glomus intraradices, under root-organ culture conditions. *New Phytologist 156* (2): 275–281.

Salazar, M., Bosch-Serra, N., Estudillos, G., and Poch, R.M. (2009). Rehabilitation of semi-arid coal mine spoil bank soils with mine residues and farm organic by-products. *Arid Land Research and Management 23* (4): 327–341.

Santos, E.S., Abreu, M.M., de Varennes, A. et al. (2013). Evaluation of chemical parameters and ecotoxicity of a soil developed on gossan following application of polyacrylates and growth of Spergularia purpurea. *Science of the Total Environment 461*: 360–370.

Santos, E.S., Magalhães, M.C.F., Abreu, M.M., and Macías, F. (2014a). Effects of organic/inorganic amendments on trace elements dispersion by leachates from sulfide-containing tailings of the São Domingos mine, Portugal. Time evaluation. *Geoderma 226*: 188–203.

Santos, E.S., Abreu, M.M., Batista, M.J. et al. (2014b). Inter-population variation on the accumulation and translocation of potentially harmful chemical elements in Cistus ladanifer L. from Brancanes, Caveira, Chança, Lousal, Neves Corvo and São Domingos mines in the Portuguese Iberian Pyrite Belt. *Journal of Soils and Sediments 14* (4): 758–772.

Santos, E.S., Abreu, M.M., Macías, F., and de Varennes, A. (2014c). Improvement of chemical and biological properties of gossan mine wastes following application of amendments and growth of Cistus ladanifer L. *Journal of Geochemical Exploration 147*: 173–181.

Santos, E.S., Abreu, M.M., Macías, F., and de Varennes, A. (2016a). Chemical quality of leachates and enzymatic activities in Technosols with gossan and sulfide wastes from the São Domingos mine. *Journal of Soils and Sediments 16* (4): 1366–1382.

Santos, E.S., Abreu, M.M., and Saraiva, J.A. (2016b). Mutielemental concentration and physiological responses of Lavandula pedunculata growing in soils developed on different mine wastes. *Environmental Pollution 213*: 43–52.

Santos, E.S., Abreu, M.M., Macías, F., and Magalhães, M.C.F. (2017). Potential environmental impact of technosols composed of gossan and sulfide-rich wastes from São Domingos mine: assay of simulated leaching. *Journal of Soils and Sediments 17* (5): 1369–1383.

Sauvé, S., Martı′ nez, C.E., McBride, M., and Hendershot, W. (2000). Adsorption of free lead (Pb2+) by pedogenic oxides, ferrihydrite, and leaf compost. *Soil Science Society of America Journal 64* (2): 595–599.

Scheunert, I., Chen, B., and Korte, F. (1989). Fate of 2, 4, 6-trichlorophenol-14C in a laboratory soil-plant system. *Chemosphere 19* (10–11): 1715–1720.

Sen, S. and Kumar, V. (2012). The effect of different soil amendments on growth of Mung bean (Vigna radiata L.) in coal mine overburden dumps. *International Journal on Emerging Technology and Advance Engineering 2* (12): 264–270.

Sengupta, M. (1993). *Environmental Impacts of Mining Monitoring, Restoration, and Control*. CRC Press.

SER (2002). *The SER Primer on Ecological Restoration*. Washington: SER.

Shakoor, M.B., Niazi, N.K., Bibi, I. et al. (2016). Remediation of arsenic-contaminated water using agricultural wastes as biosorbents. *Critical Reviews in Environmental Science and Technology 46* (5): 467–499.

Sharma, S.S. and Dietz, K.J. (2009). The relationship between metal toxicity and cellular redox imbalance. *Trends in Plant Science 14* (1): 43–50.

Sharma, P. and Pandey, S. (2014). Status of phytoremediation in world scenario. *International Journal of Environmental Bioremediation & Biodegradation 2* (4): 178–191.

Sheoran, V., Sheoran, A.S., and Poonia, P. (2010). Soil reclamation of abandoned mine land by revegetation: a review. *International Journal of Soil, Sediment and Water 3* (2): 13.

Shi, W.G., Li, H., Liu, T.X. et al. (2015). Exogenous abscisic acid alleviates zinc uptake and accumulation in *P opulus× canescens* exposed to excess zinc. *Plant, Cell & Environment 38* (1): 207–223.

Shi, X., Sun, H., Chen, Y. et al. (2016). Transcriptome sequencing and expression analysis of cadmium (Cd) transport and detoxification related genes in Cd-accumulating Salix integra. *Frontiers in Plant Science 7*: 1577.

Singh, J.S., Pandey, V.C., and Singh, D.P. (2011). Efficient soil microorganisms: a new dimension for sustainable agriculture and environmental development. *Agriculture, Ecosystems & Environment 140* (3-4): 339–353.

Sklenička, P. and Lhota, T. (2002). Landscape heterogeneity—a quantitative criterion for landscape reconstruction. *Landscape and Urban Planning 58* (2–4): 147–156.

Solís-Domínguez, F.A., Valentín-Vargas, A., Chorover, J., and Maier, R.M. (2011). Effect of arbuscular mycorrhizal fungi on plant biomass and the rhizosphere microbial community structure of mesquite grown in acidic lead/zinc mine tailings. *Science of the Total Environment 409* (6): 1009–1016.

Somers, E., Vanderleyden, J., and Srinivasan, M. (2004). Rhizosphere bacterial signalling: a love parade beneath our feet. *Critical Reviews in Microbiology 30* (4): 205–240.

Spaepen, S., Vanderleyden, J., and Remans, R. (2007). Indole-3-acetic acid in microbial and microorganism-plant signaling. *FEMS Microbiology Reviews 31* (4): 425–448.

Srivastava, N.K. and Ram, L.C. (2010). Reclamation of coal mine spoil dump through fly ash and biological amendments. *International Journal of Ecology & Development 17* (F10): 17–33.

Ssenku, J.E., Ntale, M., and Oryem-Origa, H. (2014a). The efficacy of compost, limestone and growth of Leucaena leucocephala (Lam.) de wit, Senna siamea (Lam.) and Eucalyptus grandis W. Hill ex Maid. for the restoration of bacterial functional diversity in the rhizosphere in copper tailings and pyrite soils. *Journal of Natural Sciences Research 4* (10): 39–54.

Sun, X. and Doner, H.E. (1998). Adsorption and oxidation of arsenite on goethite. *Soil Science 163* (4): 278–287.

Sun, R., Dsouza, M., Gilbert, J.A. et al. (2016). Fungal community composition in soils subjected to long-term chemical fertilization is most influenced by the type of organic matter. *Environmental Microbiology 18* (12): 5137–5150.

Sydnor, M.W. and Redente, E.F. (2002). Reclamation of high-elevation, acidic mine waste with organic amendments and topsoil. *Journal of Environmental Quality 31* (5): 1528–1537.

Tan, X.F., Liu, Y.G., Gu, Y.L. et al. (2016). Biochar-based nano-composites for the decontamination of wastewater: a review. *Bioresource Technology 212*: 318–333.

Tang, C., Chen, Y., Zhang, Q. et al. (2019). Effects of peat on plant growth and lead and zinc phytostabilization from lead-zinc mine tailing in southern China: Screening plant species resisting and accumulating metals. *Ecotoxicology and Environmental Safety* 176: 42–49. doi: 10.1016/j.ecoenv.2019.03.078.

Thangavel, P., Long, S., and Minocha, R. (2007). Changes in phytochelatins and their biosynthetic intermediates in red spruce (Picea rubens Sarg.) cell suspension cultures under cadmium and zinc stress. *Plant Cell, Tissue and Organ Culture 88* (2): 201–216.

Thijs, S., Sillen, W., Weyens, N., and Vangronsveld, J. (2017). Phytoremediation: state-of-the-art and a key role for the plant microbiome in future trends and research prospects. *International Journal of Phytoremediation 19* (1): 23–38.

Timmusk, S., Nicander, B., Granhall, U., and Tillberg, E. (1999). Cytokinin production by *Paenibacillus polymyxa*. *Soil Biology and Biochemistry 31* (13): 1847–1852.

Tisdall, J.M. (1994). Possible role of soil microorganisms in aggregation in soils. *Plant and Soil 159* (1): 115–121.

Tiwary, R.K. (2001). Environmental impact of coal mining on water regime and its management. *Water, Air, and Soil Pollution 132* (1): 185–199.

Tobin, J.M. (2001). Fungal metal biosorption. In: *British Mycological Society Symposium Series*, vol. 23, 424–444. Academic Press.

Tonin, C., Vandenkoornhuyse, P., Joner, E.J. et al. (2001). Assessment of arbuscular mycorrhizal fungi diversity in the rhizosphere of Viola calaminaria and effect of these fungi on heavy metal uptake by clover. *Mycorrhiza 10* (4): 161–168.

Topp, E., Scheunert, I., and Korte, F. (1989). Kinetics of the uptake of 14C-labeled chlorinated benzenes from soil by plants. *Ecotoxicology and Environmental Safety 17* (2): 157–166.

Tordoff, G.M., Baker, A.J.M., and Willis, A.J. (2000). Current approaches to the revegetation and reclamation of metalliferous mine wastes. *Chemosphere 41* (1–2): 219–228.

Tripti, K.A., Kumar, V., and Anshumali (2015). Effect of commercial pesticides on plant growth-promoting activities of Burkholderia sp. strain L2 isolated from rhizosphere of Lycopersicon esculentum cultivated in agricultural soil. *Toxicological & Environmental Chemistry 97* (9): 1180–1189.

Tsydenova, O. and Bengtsson, M. (2011). Chemical hazards associated with treatment of waste electrical and electronic equipment. *Waste Management 31* (1): 45–58.

Turco, R.F., Kennedy, A.C., and Jawson, M.D. (1994). Microbial indicators of soil quality. *Defining Soil Quality for a Sustainable Environment 35*: 73–90.

Turnau, K. and Haselwandter, K. (2002). Arbuscular mycorrhizal fungi, an essential component of soil microflora in ecosystem restoration. In: *Mycorrhizal Technology in Agriculture*, 137–149. Basel: Birkhäuser.

Turnau, K., Anielska, T., Ryszka, P. et al. (2008). Establishment of arbuscular mycorrhizal plants originating from xerothermic grasslands on heavy metal rich industrial wastes–new solution for waste revegetation. *Plant and Soil 305* (1): 267–280.

Turnau, K., Ostachowicz, B., Wojtczak, G. et al. (2010). Metal uptake by xerothermic plants introduced into Zn-Pb industrial wastes. *Plant and Soil 337* (1): 299–311.

Turnau, K., Gawroński, S., Ryszka, P., and Zook, D. (2012). Mycorrhizal-based phytostabilization of Zn–Pb tailings: lessons from the Trzebionka mining works (Southern Poland). In: *Bio-geo Interactions in Metal-Contaminated Soils*, 327–348. Berlin, Heidelberg: Springer.

Tuskan, G.A., Difazio, S., Jansson, S. et al. (2006). The genome of black cottonwood, Populus trichocarpa (Torr & Gray). *Science 313* (5793): 1596–1604.

Uchimiya, M., Chang, S., and Klasson, K.T. (2011). Screening biochars for heavy metal retention in soil: role of oxygen functional groups. *Journal of Hazardous Materials 190* (1–3): 432–441.

Ukwattage, N.L., Ranjith, P.G., and Bouazza, M. (2013). The use of coal combustion fly ash as a soil amendment in agricultural lands (with comments on its potential to improve food security and sequester carbon). *Fuel 109*: 400–408.

Upadhyay, S.K., Singh, D.P., and Saikia, R. (2009). Genetic diversity of plant growth promoting rhizobacteria isolated from rhizospheric soil of wheat under saline condition. *Current Microbiology 59* (5): 489–496.

US Environmental Protection Agency (2000). *Introduction to Phytoremediation*. National Risk Management Research Laboratory, EPA/600/R-99/107. http://www.clu-in.org/download/remed/introphyto.pdf (accessed date 14/09/2022).

Ussiri, D.A. and Lal, R. (2008). Method for determining coal carbon in the reclaimed minesoils contaminated with coal. *Soil Science Society of America Journal 72* (1): 231–237.

Vafadar, F., Amooaghaie, R., and Otroshy, M. (2014). Effects of plant-growth-promoting rhizobacteria and arbuscular mycorrhizal fungus on plant growth, stevioside, NPK, and chlorophyll content of Stevia rebaudiana. *Journal of Plant Interactions 9* (1): 128–136.

Van Ginneken, L., Meers, E., Guisson, R. et al. (2007). Phytoremediation for heavy metal-contaminated soils combined with bioenergy production. *Journal of Environmental Engineering and Landscape Management 15* (4): 227–236.

Vance, C.P., Uhde-Stone, C., and Allan, D.L. (2003). Phosphorus acquisition and use: critical adaptations by plants for securing a nonrenewable resource. *New Phytologist 157* (3): 423–447.

de Varennes, A., Abreu, M.M., Qu, G., and Cunha-Queda, C. (2010). Enzymatic activity of a mine soil varies according to vegetation cover and level of compost applied. *International Journal of Phytoremediation 12* (4): 371–383.

Venkateswarlu, K., Nirola, R., Kuppusamy, S. et al. (2016). Abandoned metalliferous mines: ecological impacts and potential approaches for reclamation. *Reviews in Environmental Science and Bio/Technology 15* (2): 327–354.

Venturi, V. and Keel, C. (2016). Signaling in the rhizosphere. *Trends in Plant Science 21* (3): 187–198.

Verbruggen, N., Hermans, C., and Schat, H. (2009). Molecular mechanisms of metal hyperaccumulation in plants. *New Phytologist 181* (4): 759–776.

Verma, J.P., Yadav, J., Tiwari, K.N. et al. (2010). Impact of plant growth promoting rhizobacteria on crop production. *International Journal of Agricultural Research 5* (11): 954–983.

Verret, F., Gravot, A., Auroy, P. et al. (2004). Overexpression of AtHMA4 enhances root-to-shoot translocation of zinc and cadmium and plant metal tolerance. *FEBS Letters 576* (3): 306–312.

Vessey, J.K. (2003). Plant growth promoting rhizobacteria as biofertilizers. *Plant and Soil 255* (2): 571–586.

Wang, Y., Dawson, R., Han, D. et al. (2001). Landscape ecological planning and design of degraded mining land. *Land Degradation & Development 12* (5): 449–459.

Wang, Z., Gerstein, M., and Snyder, M. (2009). RNA-Seq: a revolutionary tool for transcriptomics. *Nature Reviews Genetics 10* (1): 57–63.

Wang, Y., Zhao, Z., Deng, M. et al. (2015). Identification and functional analysis of microRNAs and their targets in Platanus acerifolia under lead (Pb) stress. *International Journal of Molecular Sciences 16* (4): 7098–7111.

Weber, J., Strączyńska, S., Kocowicz, A. et al. (2015). Properties of soil materials derived from fly ash 11 years after revegetation of post-mining excavation. *Catena 133*: 250–254.

Wiche, O., Tischler, D., Fauser, C. et al. (2017). Effects of citric acid and the siderophore desferrioxamine B (DFO-B) on the mobility of germanium and rare earth elements in soil and uptake in Phalaris arundinacea. *International Journal of Phytoremediation 19* (8): 746–754.

Wijesekara, H., Bolan, N.S., Vithanage, M. et al. (2016a). Utilization of biowaste for mine spoil rehabilitation. *Advances in Agronomy 138*: 97–173.

Wijesekara, H., Bolan, N.S., Kumarathilaka, P. et al. (2016b). Biosolids enhance mine site rehabilitation and revegetation. In: *Environmental Materials and Waste*, 45–71. Academic Press.

Wild, S.R., Berrow, M.L., McGrath, S.P., and Jones, K.C. (1992). Polynuclear aromatic hydrocarbons in crops from long-term field experiments amended with sewage sludge. *Environmental Pollution 76* (1): 25–32.

Wong, M.H. (2003). Ecological restoration of mine degraded soils, with emphasis on metal contaminated soils. *Chemosphere 50* (6): 775–780.

Wood, P.A. (1997). Remediation methods for contaminated sites. In: *Contaminated Land and its Reclamation*, 47–71. Thomas Telford Publishing.

Woś, B. and Pietrzykowski, M. (2015). Simulation of birch and pine litter influence on early stage of reclaimed soil formation process under controlled conditions. *Journal of Environmental Quality 44* (4): 1091–1098.

Wu, J., Chen, A., Peng, S. et al. (2013). Identification and application of amino acids as chelators in phytoremediation of rare earth elements lanthanum and yttrium. *Plant and Soil 373* (1): 329–338.

Wuana, R.A. and Okieimen, F.E. (2011). Heavy metals in contaminated soils: a review of sources, chemistry, risks and best available strategies for remediation. *International Scholarly Research Notices 2011*: 1–20. doi: 10.5402/2011/402647.

Wulandari, D., Cheng, W., and Tawaraya, K. (2016). Arbuscular mycorrhizal fungal inoculation improves Albizia saman and Paraserianthes falcataria growth in post-opencast coal mine field in East Kalimantan, Indonesia. *Forest Ecology and Management 376*: 67–73.

Yang, J.E., Skousen, J.G., Ok, Y.S. et al. (2006). Reclamation of abandoned coal mine waste in Korea using lime cake by-products. *Mine Water and the Environment 25* (4): 227–232.

Yang, L., Yang, Y., Chen, Z. et al. (2014). Influence of super absorbent polymer on soil water retention, seed germination and plant survivals for rocky slopes eco-engineering. *Ecological Engineering 62*: 27–32.

Yang, Y., Moore, M.J., Brockington, S.F. et al. (2015). Dissecting molecular evolution in the highly diverse plant clade Caryophyllales using transcriptome sequencing. *Molecular Biology and Evolution 32* (8): 2001–2014.

Yang, S.X., Liao, B., Yang, Z.H. et al. (2016a). Revegetation of extremely acid mine soils based on aided phytostabilization: a case study from southern China. *Science of the Total Environment 562*: 427–434.

Zhang, J., Yu, J., Hong, H. et al. (2017). Identification of heavy metal pollutant tolerance-associated genes in Avicennia marina (Forsk.) by suppression subtractive hybridization. *Marine Pollution Bulletin 119* (1): 81–91.

Zheng, P., Du, Y., and Ma, X. (2015). Selective fabrication of iron oxide particles in halloysite lumen. *Materials Chemistry and Physics 151*: 14–17.

Zhengfu, B.I.A.N., Inyang, H.I., Daniels, J.L. et al. (2010). Environmental issues from coal mining and their solutions. *Mining Science and Technology (China) 20* (2): 215–223.

Zhenqi, H., Peijun, W., and Jing, L. (2012). Ecological restoration of abandoned mine land in China. *Journal of Resources and Ecology 3* (4): 289–296.

Zhou, L., Li, Z., Liu, W. et al. (2015a). Restoration of rare earth mine areas: organic amendments and phytoremediation. *Environmental Science and Pollution Research 22* (21): 17151–17160.

Zhou, X., Jacobs, T.B., Xue, L.J. et al. (2015b). Exploiting SNP s for biallelic CRISPR mutations in the outcrossing woody perennial Populus reveals 4-coumarate: CoA ligase specificity and redundancy. *New Phytologist 208* (2): 298–301.

Zhu, J., Wei, S., Chen, M. et al. (2013). Magnetic nanocomposites for environmental remediation. *Advanced Powder Technology 24* (2): 459–467.

Zimmer, D., Baum, C., Leinweber, P. et al. (2009). Associated bacteria increase the phytoextraction of cadmium and zinc from a metal-contaminated soil by mycorrhizal willows. *International Journal of Phytoremediation 11* (2): 200–213.

Zou, J., Wang, G., Ji, J. et al. (2017). Transcriptional, physiological and cytological analysis validated the roles of some key genes linked Cd stress in Salix matsudana Koidz. *Environmental and Experimental Botany 134*: 116–129.

4

Carbon Sequestration Potential of Restored Coal Mine Soils

Abstract

Surface mining activities destroyed the land and altered geological settings and biogeochemical cycles. Ecosystem restoration is practiced worldwide to mitigate the negative impact of mining and to restore ecological integrity. Restoration of mine waste dump is often challenging because of adverse spoil conditions and overall hostile environment in post-mining sites. Overburden mine spoils lack organic matter, soil structure, and functions. In the phase of climate change, one of the major objectives of ecological restoration of mine degraded lands is to restore soil carbon pool which is largely performed through the plantation. Soil carbon accumulation improves soil quality, and hence it is essential for mine restoration. This chapter comprehensively reviews the carbon sequestration in restored coal mine soils by the means of plantation and soil amendments. Methods and challenges in quantification of soil organic carbon pool due to the presence of coal carbon were also reviewed. Carbon sequestration in reclaimed mine soil partially offsets the carbon emission from mining activities and instigates carbon cycling which helps the proper functioning of a restored ecosystem.

Keywords *Surface mining; carbon sequestration; reclaimed mine soil; plantation; soil amendments*

CONTENTS

Eco-Restoration of Mine Land, First Edition. Vimal Chandra Pandey, Jitendra Ahirwal, Roopali Roychowdhury, and Ritu Chaturvedi.

4.1 Introduction

Carbon sequestration is the process of removing atmospheric CO_2 by transferring carbon into biomass, soils, and deeper Earth. This can be performed by plants and geological processes. Plants are a reliable, natural, and economic option to store atmospheric CO_2 for the long term because they store carbon in living tissues until decomposition. The process of photosynthesis accumulates carbon in biomass, whereas the stored carbon is released into the atmosphere via root respiration (Chapin et al. 2002). Also, soil microbial respiration contributes substantially to soil carbon loss. These processes can be largely influenced by climatic conditions, soil types, and the structure of an ecosystem. Soil is the major terrestrial carbon sink that stored nearly 2157–2293 Pg carbon in the top 1 m depth (Batjes 2014). It is emphasized that increasing temperature in future climatic conditions can increase photosynthesis activities and accumulate a large amount of atmospheric CO_2 but that can be compensated by an increase in soil respiration that contributes CO_2 into the atmosphere. Therefore, developing a typical sustainable ecosystem that can act as a carbon reservoir is the prerequisite to fighting against climate change. For this, a precise assessment of the ecosystem processes related to carbon sequestration, increasing plant growth, and optimizing land use for carbon storage is important. Moreover, management of the ecosystem from the molecular level to the landscape level is essential to increasing ecosystem carbon pool.

Globally, the terrestrial ecosystems stored a large amount of carbon pool. Increasing terrestrial carbon pool by the means of natural and technological advances is an important part of an overall carbon management strategy. For this, understanding the drivers of the global carbon cycle and the carbon sequestration potential of the natural ecosystem are very much important. Global estimates reported that forests sequestered nearly 12% of anthropogenic carbon emissions

(Pan et al. 2011; Hurteau et al. 2019). Moreover, the recent technological advances in carbon capture and storage (CCS) have proven vital in mitigating greenhouse gas emissions, particularly emitted via the burning of fossil fuel, power plants, steel, cement, and other industrial application. However, the increasing rate of deforestation, land use change, and expansion of agriculture in forests have significantly reduced forest area and its carbon pool. The atmospheric CO_2 concentration has risen at an average rate of 0.33 ppm yr^{-1} from the preindustrial level to the end of the twentieth century. Nevertheless, in the last two decades, it increases at an alarming rate (2.21 ppm yr^{-1}) leading to changes in the Earth's surface temperature and global climate. Therefore, several countries are now committed to achieving sustainable management of forests and enhancement of forest carbon stocks under the ages of the $UNREDD^+$ policy. The participatory countries in the Kyoto Protocol have to monitor the national greenhouse gas emission including changes in forest carbon stock (UNFCCC 2010). For example, India committed to creating an additional carbon sink of 2.5–3.0 billion tonnes of CO_2 equivalent by enhancing its forest and tree cover by 2030 under its nationally determined contributions to the Paris Agreement.

Mining is one of the core industries vital for the economic development of a country. Not only does it contribute to GDP, but also provides raw materials to several other industries which makes it a major contributor to the national economy. Mining contributes 1.2% to the total world GDP (Ericsson and Löf 2019). For instance, mining alone can contribute 2.2–2.5% to the GDP, and in the industrial sector, it contributes about 10% in India. Mining of minerals can mainly divide into fuel and non-fuel minerals. The four major non-fuel minerals are coal, lignite, petroleum, and natural gas. Coal plays an important role in the economic development of many countries, particularly, developing countries. It is the most important fuel for energy generation and nearly 40% of the world's energy is generated by coal. Asia is the biggest market for coal accounting for 66% of global coal consumption. At the same time, most of the developed countries switch their energy generation from conventional to nonconventional energy resources like nuclear and solar energy. Though mining activities are essential for energy generation and economic growth, they cause substantial changes in the ecosystem structure and functions. Surface mining of coal has a severe environmental consequence. As most coal deposits are under the forested areas and river basins, a substantial amount of coal production through surface mining involves the destruction of land cover. Removal of entire vegetation, soil cover, and overlying rocks to mine the land results in permanent loss of soil fertility, C sink, land use change, loss of biodiversity, and ecosystem services that inherently damage and pollute the environment (Ussiri and Lal 2005; Ahirwal and Maiti 2016; Liu et al. 2017). The release of air pollutants such as CO_2, NO_x, and SO_x, and trace elements can also be an important environmental concern associated with coal mining. The inevitable changes caused by surface mining can be compensated by the restoration of post-mining land. The most appropriate techniques involved technical and biological restoration. These techniques are efficient in bringing back the state of an ecosystem before disturbance.

4.2 Generation and Management of Mine Waste

Mining generates a large amount of waste materials during mineral extraction. Removal of soils, subsoils, and overburden rocks to reach the coal seam results in the generation of waste. The inner and overburden material is removed by cutting and blasting either side of the mine. These excavation processes generated a huge amount of rock debris generally known as mine spoil. Mine spoils are generally dumped outside the quarry area by creating overburden dumps or sometimes concurrently used for backfilling. This waste often shows adverse physicochemical properties, lack of soil structure, and fertility and contains toxic elements such as heavy metals. Surface mining activities generate up to 10 times of waste compared to underground mining. Surface mining is more economic and thus commonly practiced in the countries like India where coal seams are at shallow depths. Coal production in both developed and developing countries increased tremendously (Figure 4.1). The continuous increase in coal production over the years significantly increase the generation of mine waste in major coal-producing countries. Fore example, coal is the biggest contributor to total mining waste generated in India.

Mine overburden dumps have a remarkable impact on the surrounding environment and its socioeconomic setup. Mining activities generate waste material that required large patches of land to store or manage (Maiti et al. 2016; Ahirwal and Maiti 2018). With an increase in energy production, the amount of waste material increases at a higher rate which may pose an environmental health risk (Ahirwal et al. 2016). Therefore, proper management of mine waste is required to offset its negative impacts on the environment. The overburden dumps are highly unstable, lack mechanical strength, and are prone to subsidence; therefore, best technical reclamation practices are required to stabilize these overburden dumps. Moreover, the restoration of land productivity, rehabilitation of biodiversity, and attributes of an ecosystem are challenging to restoration ecologists and land managers. Mining starts with deforestation or land clearing, removal of mine waste and generation of mine overburden dumps, excavation of coal and then backfilling of mine voids, stabilization of waste dump by revegetation of dumpsite with grass-legume followed by afforestation with selective tree species (Figure 4.2).

Revegetation of mine waste dump is essential for mine restoration. Because it not only restores the structure of an ecosystem but also brings back the biodiversity and improves the soil quality with the age of revegetation. Plant species help

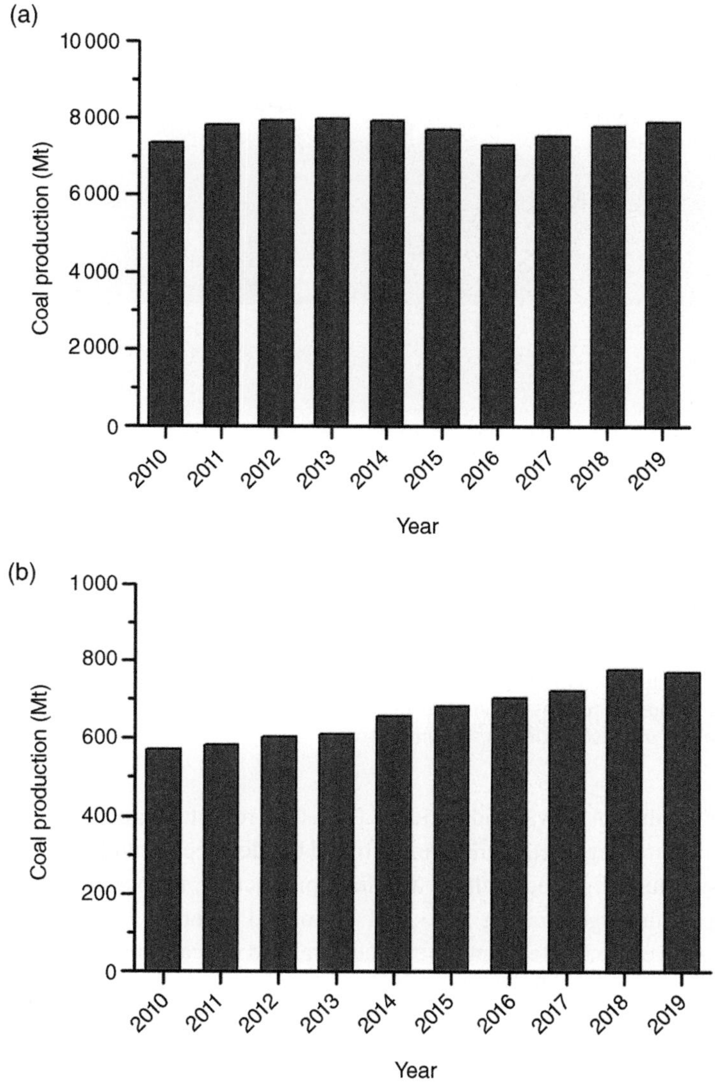

Figure 4.1 Coal production scenario of the (a) world and (b) India since 2010.
Source: Data from IEA (2020).

in land reclamation by creating favorable ecosystem conditions for natural colonization of tree species which aids in biodiversity and amelioration of soil fertility by several means such as biomass production, root, and litter decomposition, microbial growth, combating pollution, and balancing an overall ecosystem. Therefore, mining activities must ensure the development of a sustainable

Figure 4.2 Typical steps of surface mining and reclamation (a) clearing of natural forest for surface mining, (b) excavation of coal via surface mining methods, (c) revegetation of mine overburden dumps using grasses and fast-growing trees, and (d) growth of tress species along the age of revegetation in reclaimed areas.

ecosystem after mining (Ahirwal and Pandey 2020). The recent advancement in the restoration of mine degraded lands demonstrated the development of a multi-functional ecosystem such as agriculture and food production, timber and non-timber forest products, generating a pool of biological diversity, enhancing tourism, developing eco-parks, and enhancing cultural and educational values for the future and most importantly plays a role in regulating ecosystem services like carbon sequestration and climate change mitigation (Ahirwal and Maiti 2021).

4.3 Carbon Sequestration in Reclaimed Mine Soils

Ecosystem components such as plants and soil can act as "carbon sinks" by allowing more storage of C than its flow out of these systems as CO_2. Thus, there are three primary types of C sequestration systems – geological, ocean, and terrestrial ecosystems. Geological sequestration is the process of CCS in several types of geologic formations with high organic content such as depleted oil and gas reservoirs, saline aquifers, and shale formations as conventional sources. Oceans also take part in C sequestration by absorbing and storing large amounts of CO_2 from the atmosphere through its natural processes.

Terrestrial carbon sequestration transforms atmospheric CO_2 into biomass components through photosynthesis and later on its decomposition transfers into the soils in the form of soil organic matter (SOM) and humus. That results in long-term storage of atmospheric CO_2 into biomass components until decomposition. Soils contain approximately 75% of the terrestrial C pool – threefold greater than that stored in biomass and hence play a major role in the global C cycle (Ussiri and Lal 2005). Various strategies have been implemented with current technologies to store additional carbon in terrestrial ecosystems (e.g. forests, agricultural lands, and wetlands) where they are maintained, enhanced, or manipulated to increase their ability to store C (IPCC 2000). Terrestrial sequestration can be implemented by the management of agricultural and forested systems such as trapping of C in solid form (wood products, timber, plywood, and other building materials from trees). The aboveground living biomass has a significant contribution toward an increase in C sequestration where mined sites have been reclaimed to the forest. This contribution also depends on the type of forest and species.

SOM is an important component of the soil as it influences various soil properties and is essential for nutrient availability to plants and microorganism. The C content of SOM ranged from 40 to 60%, therefore, it plays a vital role in the global C cycle (Palmer et al. 2002). The status of SOM has often been used to determine the impact of anthropogenic or natural disturbance on soil quality (Palmer et al. 2002). Surface mined land has been subjected to various physicochemical and biological alterations such as a change in soil structure and loss of organic matter (OM) and nutrients, which altogether affect the mine soil functions. The loss of SOM during the surface mining activities and their accretion during the restoration are vital to maintaining the fertility of the mine soil. There has been a massive loss of soil organic carbon (SOC) during the topsoil removal and stockpiling during the surface mining which may be due to the mineralization of the SOM and lack of fresh C input (Banning et al. 2008). Studies have shown that estimation of SOC is generally used as a proxy for SOM determination; however, the presence of the coal C often overestimates the SOC concentration in the coal mine soils (Ingram et al. 2005; Ussiri et al. 2014).

The potential of SOC sequestration in soils can be regulated by the complex interaction of different SOC pools with the microbial community and the biotic and abiotic factors over time. In general, carbon supplied by aboveground biomass, root decomposition, litter decomposition, mycorrhizal association, wind or water deposition, and other carbon managements activities such as application of biochar, woodchips, manure, and other biowaste is important for soil fertility. In contrast, the process of carbon output from the soils can be controlled by oxidation, mineralization, soil erosion, and leaching processes. It is largely believed that climate change accelerates the decomposition of carbon because an increase in temperature enhances the enzymatic activity related to decomposition that in turn enhances the emission of CO_2 into the atmosphere and leads to global warming in the future. Also, plant aboveground biomass is the major contributor to stable OM, but recent studies reported that plant root and fungal necromass are the major contributors to the stable OM. It is reported that SOC has a residence time between 1 month and 1 year, whereas the carbon associated with clay minerals has a residence time of over 100 years (Blanco-canqui and Lal 2004). Authors also opined that the adsorption of organic molecules by clay minerals in association with electrostatic forces forms an organo-mineral complex. The organo-mineral complex sequestered carbon via complex physical and biochemical interaction; organic material interacts with clay particles via adsorption, clay surface polymerizes humic substance, and these polymerized organic compounds are physiochemically sequestered by clay particles that are inaccessible to soil microbes thus retaining it for a longer time (Laird et al. 2001).

Several strategies such as plantation and amendments with waste have been developed to increase SOC in the mined land, but most of the time adverse mine soil properties retard the plant growth (Ahirwal et al. 2020). Estimation of the SOC in reclaimed mine soil (RMS) with the integrated biotechnological approach showed a sevenfold increase in that SOC over the 20 years (Juwarkar et al. 2009). While investigating the SOM dynamics in lignite-rich mine soils of Germany under a chronosequence of pine, Rumpel et al. (1999) found an increase in recent C accumulation from 11 years ($1.28\,kg\,m^{-2}$) to 32 years ($5.02\,kg\,m^{-2}$) in the top 5-cm profile, which is much higher than the natural forest soil ($3.82\,kg\,m^{-2}$). With time, the development of vegetation cover supplies fresh C to mine soil that is later decomposed by the microorganisms and increases the mine soil fertility. Mine soils often showed a low SOC concentration that promises a high SOC sequestration potential with time which can be a sensitive indicator of the mine soil quality. Ingram et al. (2005) reported that the minimum concentration of 0.52% SOC is essential to maintain the nutrient cycle in the RMS.

Tropical forests account for 40–50% of the C in terrestrial biomass (Nyirambangutse et al. 2017). They provide important ecosystem services like C accumulation; however, surface mining significantly reduced the ecosystem services by removing plant and soil cover. Therefore, assessment of C stock in revegetated coal mine land is important to balance the global C equation, and the regenerated ecosystem in the post-mining land has different soil quality and ecosystem functions. It can be observed that C accumulation in RMS is increased with the age of revegetation. The estimate of SOC stocks in the deep soil is different in many regions of the world. Investigation of the influence of parent materials on soil C stocks reveals that SOC stocks in topsoil (0–30 cm) were much higher in forests ($83.4\,Mg\,C\,ha^{-1}$) than in the cultivated soils ($48.9\,Mg\,C\,ha^{-1}$) (Barre et al. 2017).

In a terrestrial ecosystem, soil C stock is usually greater than the living vegetation and atmosphere. Therefore, it is important to understand the role of SOC in the global C cycle, early stages of pedogenesis, and ecological restoration of mine soils (Post and Kwon 2000). The C balance of an ecosystem may be defined as the difference between the amount of C gain and losses at a point in time. C accumulated in the form of complex molecules via photosynthesis, and it is lost by the process of soil respiration (Heimann and Reichstein 2008). The total mass of C stored in undisturbed soil indicates the net balance between the rate of C input and mineralization (Post and Kwon 2000). However, mine soil is drastically disturbed soil which lacks these nutrient cycling processes till they are reclaimed or restored properly. During surface mining, ecosystem SOC stock is severely depleted due to the complete removal of vegetation and topsoil cover. Once degraded lands are reclaimed/restored, the recovery of SOC stock is initially very rapid and reaches equilibrium with time. Over the years, numerous studies have been carried out in post-mining sites in many countries of the world to estimate C sequestration potential of RMS (Table 4.1). It can be observed that SOC and its sequestration rate varied at a larger scale and ranged between 0.1 and 5.2 Mg C ha^{-1} yr^{-1}.

A meta-analysis of SOC accumulation in opencast coal and oil shale mining in the northern hemisphere revealed that SOC accumulation was not related to vegetation types (Vinduskova and Frouz 2013). Authors also reported that the rate of SOC accumulation was maximum after 5–10 years and then decreased in deciduous forests. The rate of SOC accumulation was negatively correlated with the temperature under conifer forests and positively related to the grassland. Asensio

Table 4.1 Soil organic carbon stock (Mg C ha^{-1}) in afforested coal mine site.

Mine type	Country	Age of afforestation (yr)	Soil depth (cm)	SOC stock (Mg C ha^{-1})	Authors
Coal	China	11	0–30	32.6	Yan et al. (2020b)
Coal	India	16	0–20	7.96	Ahirwal and Maiti (2018)
Coal	USA	25	0–15	107.0	Shrestha and Lal (2010)
Coal	China	25	0–20	96.1	Yuan et al. (2017)
Coal	India	16	0–15	9.82	Mukhopadhyay and Masto (2016)
Coal	USA	70	0–15	45.5	Wang et al. (2016)
Coal	USA	15	0–30	24.13	Jacinthe and Lal (2007)
Coal	India	5	0–20	11.12	Singh et al. (2006)

et al. (2014) studied the effect of reclamation on soil C fraction in mine soils of Spain and found significant increases in SOC concentration from 1.4–6.6 to 10–112 g kg^{-1} due to vegetation and waste amendments, respectively. The increase or decrease in SOC concentration along the age of revegetation may depend on the supply of fresh C, plant growth, and plant traits. Studies carried out in tropical, arid, and semi-arid conditions have shown a continuous increase in SOC concentration up to more than 20 years of plant growth (Kumar et al. 2015; Mukhopadhyay and Masto 2016; Ahirwal et al. 2018; Yuan et al. 2018a).

Numerous studies recommended the use of waste materials for mine site reclamation and also emphasized that waste was the only treatment that increases the humified SOC. The effect of various tree species and soil conditions on C sequestration was studied in reforested post-mining sites and found that higher microbial activity supports the C sequestration. In different plant species mixtures, soil C stocks ranged between 3 and 12 Mg C ha^{-1} (Jozefowska et al. 2017). Assessment of C dynamics in a chronosequence of 4–13 years of reclaimed land under agriculture use in Shandong Province, China, reported a gradual increase in SOC concentration along with the age of reclamation until it reached up to 7.98 g C kg^{-1} and showed improvement in RMS quality (Qu et al. 2017). RMS C pool under grass and forest lands stored 29.7 Mg C ha^{-1} after 30 years compared to 9.11 Mg C ha^{-1} in 1 year (Chatterjee et al. 2009). Similarly, mine soil under forest and grassland cover sequestered C at a rate of 0.7 and 0.8 Mg C ha^{-1} yr^{-1} after 15 and 20 years, respectively (Akala and Lal 2001). C storage under *Dalbergia sissoo* plantation on reclaimed mine land in India was linearly increased with the age of reclamation, where mine soils accumulate 1.19 Mg C ha^{-1} after 2 years and 9.82 Mg C ha^{-1} after 16 years of reclamation (Mukhopadhyay and Masto 2016). After 19 years of revegetation with mixed plantation in Singrauli, India, RMS sequestered 22.9 Mg C ha^{-1} at a rate of 1.2 Mg C ha^{-1} yr^{-1} (Tripathi et al. 2014). The differences among the study sites may be attributable to a difference in plant species and SOM content (Frouz et al. 2009).

Most of the studies reported SOC stock in reclaimed coal mine land; however, accurate estimation of SOC stock is often a topic of debate in mine soils. Therefore, some of the studies carried out in mine soils reported different organic C fractions such as labile C, dissolve organic C, and recalcitrant C fraction which may provide accurate estimates of soil C. For example, the estimates of labile C fractions in three different monoculture plantations at a reclaimed coal mine in China reported an increase in easily oxidizable carbon (EOC) stock from 0.37 to 4.95 Mg C ha^{-1} under *Robinia pseudoacacia*, from 0.37 to 2.84 Mg C ha^{-1} under *Pinus tabuliformis*, and from 0.37 to 2.78 Mg C ha^{-1} under *Ulmus pumila* (Yuan et al. 2018b). The study of C fractions in mine soils in India shows that RMS store 78.4 g C kg^{-1} that contain a significant amount of coal C (34.9 g C kg^{-1}), inorganic C (5.3 g C kg^{-1}), labile C (9.2 g C kg^{-1}), and recalcitrant C (24.2 g C kg^{-1}) (Das and

Maiti 2016). Similarly, a chronosequence study of reclaimed coal mine soil in West Virginia shows 8–20 g kg^{-1}, of which 7–43% was accounted for humin fraction, and labile C was reported in the range of 4.5–9.8 g C kg^{-1}. The contribution of different C fractions in total organic carbon (TOC) emphasizes the need for C fractionation study in reclaimed coal mine soils.

4.4 Carbon Fractionation: Importance and Challenges in Coal Mining Areas

SOM contains different carbon fraction which mainly includes organic and inorganic carbon of different stability. Moreover, in mine soils, geogenic carbon or coal carbon share a significant part of the OM. Thus, organic, inorganic, and coal carbon are the major carbon fraction of the mine soils (Figure 4.3). The quantification of SOM in mining areas has been receiving large attention because of its role in the global C cycle and C sequestration. Therefore, the precise estimation of SOM is important to understand C dynamics in the soil. The traditional methods

Figure 4.3 Types of carbon pool in reclaimed mine soil. Organic carbon contributed by soil microfauna, biomass, and litter decomposition constitutes 50% of the total carbon. Inorganic carbon constitutes 8–15% of total soil carbon and mainly contributed by calcite and dolomite, and coal carbon, which is geogenic in nature, comes from coal particles and contributes about 30–40% to the total carbon pool.

for determining SOM in forest/agriculture soil have been extensively used for the mine soil. Unanimously, loss on ignition (LOI) and dry combustion by the elemental analyzer are the most widely used methods for the quantification of OM and C in the mine soils. LOI method involves the combustion of the soil sample in a muffle furnace to destroy all OM, and after the combustion lost mass is assumed to be OM. With the advances in analytical techniques, several indirect methods have been developed for characterizing SOM which is relatively rapid and inexpensive. For the precise quantification of SOM, many authors used different temperatures for the combustion of soil samples. Ball (1964) found greater accuracy at 350 °C for overnight compared to 850 °C for the estimation of SOM in the non-calcareous soil. They also suggest that at 375 °C incomplete combustions were observed; however, structural water loss at high temperature (>800 °C) will increase the total mass loss resulting in an overestimation of SOM. Hoogsteen et al. (2015) and De Vos et al. (2005) observed that combustion at 550 °C for 3 hours can remove all the SOM. While estimating the SOM by the LOI methods, operating parameters like temperature, soil mass, duration of combustion, types of furnace, and arrangement of the samples during combustion may underestimate or overestimate the SOM content; therefore, it needs to be standardized for the specific soil types.

4.4.1 Physical Fractionation

Physical fractionation methods are used to differentiate C fractions based on structural and functional complexity in soil primarily based on the size of the OM (Jagadamma and Lal 2010):

i) Particulate organic matter (POM): it is a primary source of energy for microorganisms and constitutes a labile C pool. It can contribute up to 40% to total SOM in soils permanently covered by vegetation. The turnover time of POM is slow compared to recently added litter but faster than that of mineral associate SOC (Jagadamma and Lal 2010).
ii) Mineral associate organic matter (MAOM): it is defined as OM bound to mineral particles. SOM associated with <2 μm particle-size fractions of soils has slow turnover times, and it is assumed to be stabilized mainly by interaction with minerals (Chenu and Plante 2006). These fractions have a very slow turnover time and are generally referred to as humic substances (Hassink et al. 1997).

Estimation of SOC fractions based on their physical characteristics is reported by several researchers (Jagadamma and Lal 2010; Song et al. 2012). Wet sieving methods are generally used to distinguish sand size (>53 μm), silt size (2–53 μm), and clay size (200 μm) by using various dispersion reagents such as $MgCl_2$, NaI, and $CaCl_2$. The light fractions with low density (<1.7 g cm^{-3}) and heavy fractions with high density (>1.7 $g\,cm^{-3}$) are referred to decayed plant and animal products and humic substances, respectively. After separating the various C fraction, it is analyzed for C concentration using TOC and/or carbon–hydrogen–nitrogen–sulfur (CHNS) elemental analyzer.

4.4.2 Chemical/Thermal Oxidation Methods

4.4.2.1 Cold and Hot Water Extractable Carbon

Water-soluble extractable (WSE) or cold water extractable C is the C portion that can be rapidly utilized and readily available as a respiratory C substrate. It contains very little C derived from the microbial cell and is very less compared to the hot water extractable (HWE) C pool. HWE C is referred to as a pool of mineralizable SOM which is partially derived from soil microbial C and represents a readily available source of nitrogen in the soil (Ghani et al. 2003). Presently, we reviewed only the methods to determine WSE and HWE C in mine soils and forest soils as described in Ghani et al. (2003).

i) Water-soluble extract (WSE) C: Field fresh samples of <2 mm size were used for C estimation. Oven-dried soil samples equivalent to 3 g were transferred to polypropylene centrifuge tubes. The sample was extracted with 30 mL of distilled water for 30 minutes in a shaker water bath at 30 rpm for 20 °C. Samples were centrifuged for 20 minutes at 3500 rpm and then the supernatant was filtered through 0.45 μm filter paper for the analysis in the TOC analyzer.
ii) Hot-water soluble extract (HWS) C: A further 30 mL of distilled water was added to the remaining soil samples in the same tube. This tube was shaken in a mechanical shaker for 1 minute to suspend the soil in water. The centrifuge tube was left in a shaker water bath at 80 °C for 16 hours. After the end of this period, the tube was again shaken for 1 minute to suspend all the OM in an extraction medium. This tube was centrifuged at 3500 rpm for 20 minutes. The supernatant was filtered through a 45 μm cellulose nitrate membrane filter. Filtered samples were analyzed for hot water extractable carbon (HWC) in the TOC analyzer

4.4.2.2 Acid Hydrolysis/Sequential Extraction Method

The sequential extraction method distinguishes soil inorganic C (SIC), SOC, and fossil C in the mine soil samples (Ussiri and Lal 2008; Das and Maiti 2016). We represent the three-step methods to differentiate various C fractions. Acid (HCl) treatment removes the carbonates bound to soils, and alkali and acid treatments remove the base hydrolyzable C and acid hydrolyzable C, respectively. Treatment with hydrofluoric acid (HF) removes mineral-bound C, and fossil C was determined by the ignition method.

Total C content is determine on oven-dried soil sieving to <250 μm and determine by the dry-combustion method. SIC, SOC, and fossil C are determine by the sequential-extraction method (Ussiri and Lal 2008; Das and Maiti 2016). For SIC, soil samples were treated with 1 M HCl in a 1 : 10 ratio (soil : HCl, w/v) for 30 minutes to remove carbonates, and residues were recovered by centrifugation (3000 rpm for 10 minutes). The difference between the total C and C content of the residues was considered the SIC. For SOC, carbonate-free soil was dried at 60 °C, treated with 0.5 M NaOH (1 : 10, soil : NaOH, w/v), and allowed to react in an orbital shaker at 100 rpm for 1 hour. Samples were allowed to react for 15 hours and centrifuged at 3000 rpm, and the supernatant was discarded. The residual sediments were washed twice with deionized water to extract all the OM and excess NaOH. The recovered sediments were dried and allowed to react with 60% HNO_3 (1 : 10, soil : HNO_3, w/v) in an orbital shaker at 100 rpm for 15 minutes. Samples were centrifuged, and the supernatant was discarded. To remove mineral-bound C, samples were treated with 10% HF (1 : 10, soil : HF, w/v) and allowed to react for 4 hours with occasional manual stirring. Sediments were again reacted with 20 mL of 0.5 M NaOH (1 : 10, soil : NaOH, w/v) for 30 minutes. Sediments were recovered by centrifugation, and samples were washed with deionized water, dried at 60 °C for 48 hours, and analyzed for total C using the elemental analyzer. The remaining sediments after the sequential extraction process were placed in a preweighted crucible and ignited at 350 °C for 3 hours to remove highly recalcitrant C. For fossil C, the sediments remaining after the ignition process were analyzed for total C by the elemental analyzer.

4.4.3 Biological Stability of Soil Carbon

To determine the labile and recalcitrant pool of SOC, we detailed the method described by Rovira and Vallejo (2002):

4.4.3.1 Labile Pool

The labile pool represents a part of recent C derived from microorganisms and dead plant roots. It takes part in the mineralization process and provides nutrients to plants. Recalcitrant C pools are a stable fraction of C that cannot easily take part in the mineralization process and decay resistance. Labile C is a part of OM that can be readily decomposed by soil organisms. It can be determined by the chemical treatment (Blair et al. 1995; Vieira et al. 2007). In this method, two concentrations of $KMnO_4$ (333 and/or 60 mM) were used to oxidize organic C present in the soils. For this, 15 mg of soils were placed in a 100 mL screw-cap tube and added 25 mL $KMnO_4$ and suspensions were horizontally shaken at 60 rpm for 1 hour. Then samples were extracted by centrifugation at 715 g for 5 minutes. The supernatant solution was separated and diluted with distilled water (1 : 1000 for 333 mM $KMnO_4$ and 1 : 250 for 60 mM $KMnO_4$). The solution was quantified by a spectrophotometer at 565 nm. The depletion of $KMnO_4$ concentration was directly proportional to the concentration of oxidizable C (labile C). It can be assumed that 1 mM of $KMnO_4$ is consumed to oxidize 9 mg of C. To avoid the photo-oxidation of $KMnO_4$, care was taken to avoid incidence with direct light. This method can provide the estimates of the chemically oxidizable C present in soils.

4.4.3.2 Humified Organic Matter

Humified OM consists of humin, fulvic, and humic acids which have different chemical stability and degree of humification. The chemical method described by de Blas et al. (2010) and Asensio et al. (2014) can distinguish the humified OM fractions. For this, free organic matter (FOM) was removed by treating soils with the 2 M H_3PO_4 followed by centrifugation and filtration. The extract recovered after FOM analysis (soluble extract I) consisted of phosphoric acid extracted fulvic acid fraction (FAP). The residue after FOM extraction (residue I) was successively extracted with 0.1 M $Na_4P_2O_7$ before the extraction of humic acid C (HA) and fulvic acid C (FA), as NaOH can strongly react with soil and may give erroneous results. After extraction, samples were under successive extractions with 0.1 M NaOH to extract all the organic C. Supernatants from $Na_4P_2O_7$ and NaOH were stored together as a homogeneous sample (soluble extract II). TOC concentration of the final sample (extract) is the extractable C (C_{ext}). Part of this combined sample was saved, and H_2SO_4 was added to allow the humic acids to precipitate. The supernatant after this precipitation was analyzed in the TOC to determine C. The difference between C_{ext} and C_{FA} is the C_{HA}. The residue after the last extraction with NaOH (residue II) corresponds to the humin fraction, which was dried and ground. Its organic C was determined in a solid module coupled with a TOC analyzer.

4.5 Carbon Indices

Carbon indices have been used to assess the C accumulation in the soil of different land uses (Blair et al. 1995; Rovira and Vallejo 2007; Vieira et al. 2007; Silveira et al. 2008). Various indices like recalcitrant index, carbon pool index (CPI), liability index, and carbon management index (CMI) were used to evaluate C accumulation in RMS. Recalcitrant index (RI_C) was calculated as the percent ratio of non-hydrolyzable C (recalcitrant C) to that of total organic C. It provides an estimate of the recalcitrant C pool in the total C pool of that sample:

$$\text{Recalcitrant index} = \frac{\text{Unhydrolyzable C}}{\text{Total organic C}} \times 100 \tag{4.1}$$

The CPI was calculated as the ratio of TOC in mine soils to that of TOC in control forest soil and gives an estimate of the total C pool:

$$\text{Carbon pool index}\left(\text{CPI}\right) = \frac{\text{TOC in reclaimed mine soil}}{\text{TOC in forest soil}} \tag{4.2}$$

The lability index (LI) indicates the status of the POM in the soils:

$$\text{Lability of carbon} = \frac{\text{KMnO}_4\ \text{oxidizable C}}{\text{Non}-\text{labile C}} \tag{4.3}$$

$$\text{Liability index}\left(\text{LI}\right) = \frac{\text{Lability of C in RMS}}{\text{lability of C in forest soil}} \tag{4.4}$$

The CMI is derived from the C pool and C lability. It can be a sensitive indicator of the impact of reclamation techniques on changes in SOC pool and its accumulation rate in post-mining land.

$$\text{Carbon management index} = \text{CPI} \times \text{LI} \times 100 \tag{4.5}$$

4.6 Mine Soil Amendments

Mine soils have many artifacts and are highly modified due to applied reclamation techniques thus known as Technosols (IUSS-WRB 2015). Reconstruction of mine degraded lands through technical and biological restoration practices potentially transforms the mine spoil into some usable form. Mine spoil lacks soil fertility; therefore, numerous physical, chemical, and biological amendments can be used to restore soil fertility. Although the application of topsoil is widely recognized to restore mine soil fertility, the quality of topsoil can be varied largely across the climate and land use which may not contain enough fertility components. Therefore, various soil amendments were used to enhance mine soil fertility and its carbon stock. For instance, the application of Yellow lupin green manure cropping for carbon enrichment in the sand mine cast of Poland showed an accumulation of 13.55 Mg C ha^{-1} after 1 year of reclamation (Pietrzykowski et al. 2017). Use of municipal sewage sludge for 17 years increases SOC concentration up to 115% in limestone mine soils (Ojeda et al. 2015). Moreover, SOC concentration in reclaimed mine soils was increased threefold after biochar addition (Ghosh et al. 2020).

The uses of conventional and nonconventional soil amendments such as chemical fertilizers and manure, selective plantation, and advanced physicochemical treatments help restore the mine soil carbon stock (Table 4.2). For instance, the

Table 4.2 Application of conventional and nonconventional soil amendments and selective plantation to enhance organic carbon storage in restored mine soil.

Study site	Mine types	Amendment types	Changes in SOC stock (Mg C ha^{-1})	Time since amendments	Authors
Soil amendment					
Iberian Peninsula, Spain	Limestone mine	Sewage sludge	9.0–28.0	10 years	Carabassa et al. (2020)
Loess Plateau, China	Coal mine	Chicken manure	12.51–26.32	10 years	Huang et al. (2019)
		Chemical fertilizer, chicken manure	12.51–24.70		
Central Poland	Coal mine	Sewage sludge	12.0–33.0	18 months	Placek-Lapaj et al. (2019)
		Compost, lake chalk	12.0–45.0		
British Columbia, Canada	Copper and molybdenum mines	Biosolids	23.0–155.0	13 years	Antonelli et al. (2018)
Southeast Spain	Marble mines	Topsoil, marble sludge	5.49–11.22	6 years	Simón et al. (2018)
Selective plantation					
Jharkhand, India	Coal mine	*Stylosanthes hamata,*	2.14–3.15	2 years	Kumari and Maiti (2019)
		Cenchrus ciliaris	2.14–2.54		
Loess Plateau, China	Coal mine	*Platycladus orientalis,*	3.19–11.56	14 years	Yan et al. (2020a)
		Rhus typhina	3.19–8.69		
Telangana, India	Coal mine	*Prosopis juliflora*	1.05–21.20	8 years	Ahirwal et al. (2018)
Shanxi Province, China	Coal mine	*Robinia pseudoacacia*	1.48–12.92	18 years	Yuan et al. (2018a)
		Pinus tabuliformis	1.48–14.04	25 years	
		Ulmus pumila	1.48–6.49	20 years	
Jharkhand, India	Coal mine	*Dalbergia sissoo*	1.19–9.82	16 years	Mukhopadhyay and Masto (2016)

synergistic effect of biochar along with compost and AMF in post-mine sandpit showed an increase in SOC concentration from 1.18 to 2.19 g kg^{-1} after one year of incubation (Ohsowski et al. 2018). Muñoz-Rojas et al. (2018) analyzed the effect of N-fixing cyanobacteria inoculated with soil used in the restoration of post-mining sites and found a threefold (0.6–1.9 g kg^{-1}) increase in SOC concentrations. The addition of fly ash in combination with farmyard manure, chicken litter, plant litter, and biochar to mine spoil showed lower cumulative CO_2 emission and higher stable carbon concentration compared to without fly ash added to mine spoil after one year of incubation (Mukhopadhyay and Masto 2019). These advanced restoration practices significantly increase the mine soil carbon stocks and also help increase mine soil fertility in the long term.

4.7 Carbon Sequestration in Revegetated Coal Mine Soils: A Chronosequence Approach

Surface coal mining activities inevitably result in deforestation and land degradation. The land losses its vegetation cover, fertile topsoil, and associated soil nutrients. Over time, these coal mine spoils show unfavorable soil physicochemical properties and nutritionally poor conditions that restrict plant growth. The area can partially support the natural colonization of various species and often remain degraded/unfertile until reclaimed by forestry or by other reclamation practices. A study was carried out in reclaimed coal mine dumps of Singareni Collieries Company Limited, Telangana, India, to assess the role of selective plantation in the establishment of vegetation cover on the reclaimed coal mine land and nutrient stocks over the eight years of revegetation (Ahirwal et al. 2017). The area experienced a hot and dry climate with a mean annual temperature of 27 °C and received an average of 975 mm rainfall per year. The soil was predominantly of a sandy loam texture, and no vegetation cover was observed before the plantation. The study aimed to assess the role of selective plantation in the accretion of SOC stock of the revegetated coal mine land. Post-mining land was afforested with *Prosopis juliflora*. Afforestation was carried out by digging 2000 plantation pits (30 cm × 30 cm × 30 cm) per hectare of land, and nursery growth plant saplings were planted in each pit with the mixture of topsoil and loose dump in a ratio of 1 : 4 (w/w). A chronosequence study was carried out continuously for eight years to access changes in SOC stock.

Prosopis juliflora is a xerophytic shrub to small trees that can tolerate a wide range of environmental conditions such as alkaline soils, wide temperature range, and adverse soil conditions. Afforested coal mine dumps of age one to eight years and an undisturbed *Tectona grandis* forest were selected for the study. Mine soil samples were randomly collected from the afforested sites of different ages (eight dumps) and natural forests at 0–20, 20–40, and 40–60 cm depth by laying five quadrates (10 m × 0 m) in each dump. Replicate samples were collected from each depth. The samples were air-dried and passed through a 2-mm sieve to remove rock debris and coarse root and litter. The fine-earth samples were analyzed for SOC, and SOC stocks were calculated using the bulk density of fine-earth materials and organic carbon concentration in the particular soil depth.

The loss of plant cover and fertile topsoil due to surface mining activities resulted in the loss of SOC. Post-mining lands were reclaimed by technical reclamation techniques and then afforested with the selective plantation that helps amelioration of soil quality and increase in SOC concentration during eight years of afforestation. SOC concentration (0–20 cm) in afforested mine soils increased with the age of afforestation, and after eight years, it was 0.80 and 1.28% in the reference forest site. In the present study, SOC concentration was low at the initial stage of afforestation mainly due to the presence of mine spoil and the absence of vegetation cover; however, along the age of afforestation, SOC was accumulated on the top surface as well as in the subsoil that resulted in a significant increase in SOC concentration. The vertical distribution of the SOC concentration for all the investigated soil depths (0–20, 20–40, and 40–60 cm) was significantly different being highest in the top layer and decreasing with the soil depth.

The SOC stock in afforested mine soil increased from 20.2 to 45.3 Mg C ha^{-1} in the top 60 cm depth along with the chronosequence age. The SOC stock at the reference forest site was 30.6 Mg C ha^{-1} in 0–20 cm depth and 66.7 Mg C ha^{-1} in 0–60 cm soil depth (Figure 4.4a). The recovery of SOC stock was computed with respect to the SOC stock in the reference forest site which shows that after eight years of revegetation, SOC stocks were recovered by 68% (Figure 4.4b). The SOC stock increases along with the age of afforestation. The differences in SOC stocks along the soil depth demonstrated that the top layer exhibited 46% of the total SOC after eight years of afforestation. Development of site characteristics and attributes of an ecosystem similar to the reference site or the site before the disturbance is the main aim of restoration. Recovery of SOC stock is vital for performing several ecosystem functions including enhancing soil fertility and promoting the establishment of plant species.

4.8 Conclusion

Overall, it can be concluded that restored mine soil provides a unique opportunity to sequester atmospheric carbon and help reduce the increased amount of atmospheric carbon concentration. These restored mine sites not only play an important role in carbon sequestration but also help in increasing mine soil fertility, support plant growth, and act as a platform for microbial growth. The carbon fraction in

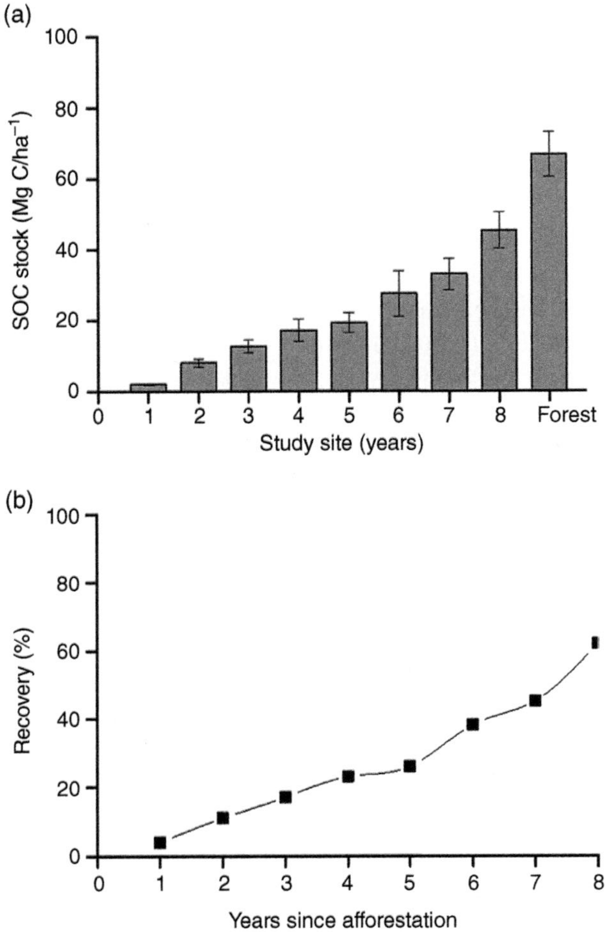

Figure 4.4 (a) Soil organic carbon stock (0–60 cm) in restored coal mine soils of one to eight years and (b) recovery of soil organic carbon along the age of afforestation compared to natural forest.

restored mine soils, particularly in coal mine soil, shows a wide range of stability that can be analyzed by different methods and separated based on its physics, thermal, chemical, and biological stability. Thus, estimation of carbon stock in restored mine soils should be based on the different carbon fractions. It is recommended that long-term research should be carried out to assess the potential of other restoration practices to increase carbon sequestration in mine soils. The existing restoration techniques may not be suited to the rapidly changing environment; therefore, the use of restoration techniques that can be sustained in future climatic conditions should be tested and adopted for mitigating environmental pollution and increasing carbon sequestration.

References

Ahirwal, J. and Maiti, S.K. (2016). Assessment of soil properties of different land uses generated due to surface coal mining activities in tropical Sal (Shorea robusta) forest, India. *Catena* 140: 155–163.

Ahirwal, J. and Maiti, S.K. (2018). Development of Technosol properties and recovery of carbon stock after 16 years of revegetation on coal mine degraded lands, India. *Catena* 166: 114–123.

Ahirwal, J. and Maiti, S.K. (2021). Restoring coal mine degraded lands in India for achieving the United Nations-Sustainable Development Goals. *Restoration Ecology* 30: e13606: https://doi.org/10.1111/rec.13606.

Ahirwal, J. and Pandey, V.C. (2020). Restoration of mine degraded land for sustainable environmental development. *Restoration Ecology* 29: e13268.

Ahirwal, J., Maiti, S.K., and Singh, A.K. (2016). Ecological restoration of coal mine-degraded lands in dry tropical climate: what has been done and what needs to be done? *Environmental Quality Management* 26: 25–36.

Ahirwal, J., Maiti, S.K., and Reddy, M.S. (2017). Development of carbon, nitrogen and phosphate stocks of reclaimed coal mine soil within 8 years after forestation with Prosopis juliflora (Sw.) Dc. *Catena* 156: 42–50.

Ahirwal, J., Kumar, A., Pietrzykowski, M., and Maiti, S.K. (2018). Reclamation of coal mine spoil and its effect on Technosol quality and carbon sequestration: a case study from India. *Environmental Science and Pollution Research* 25: 27992–28003.

Ahirwal, J., Kumar, A., and Maiti, S.K. (2020). Effect of fast-growing trees on soil properties and carbon storage in an afforested coal mine land (India). *Minerals* 10: 840.

Akala, V.A. and Lal, R. (2001). Soil organic carbon pools and sequestration rates in reclaimed mine soils in Ohio. *Journal of Environmental Quality* 30: 2098–2104.

Antonelli, P.M., Fraser, L.H., Gardner, W.C. et al. (2018). Long term carbon sequestration potential of biosolids-amended copper and molybdenum mine tailings following mine site reclamation. *Ecological Engineering* 117: 38–49.

Asensio, V., Vega, F.A., and Covelo, E.F. (2014). Effect of soil reclamation process on soil C fractions. *Chemosphere* 95: 511–518.

Ball, D.F. (1964). Loss-on-ignition as an estimate of organic matter and organic carbon in non-calcareous soils. *European Journal of Soil Science* 15: 84–92.

Banning, N.C., Grant, C.D., Jones, D.L., and Murphy, D.V. (2008). Recovery of soil organic matter, organic matter turnover and nitrogen cycling in a post-mining forest rehabilitation chronosequence. *Soil Biology and Biochemistry* 40: 2021–2031.

Barre, P., Durand, H., Chenu, C. et al. (2017). Geological control of soil organic carbon and nitrogen stocks at the landscape scale. *Geoderma* 285: 50–56.

Batjes, N.H. (2014). Total carbon and nitrogen in the soils of the world. *European journal of soil science* 65 (1): 10–21.

Blair, G.J., Lefroy, R.D., and Lisle, L. (1995). Soil carbon fractions based on their degree of oxidation, and the development of a carbon management index for agricultural systems. *Australian Journal of Agricultural Research* 46: 1459–1466.

Blanco-Canqui, H. and Lal, R. (2004). Mechanisms of carbon sequestration in soil aggregates. *Critical Reviews in Plant Sciences* 23 (6): 481–504.

de Blas, E., Rodríguez-Alleres, M., and Almendros, G. (2010). Speciation of lipid and humic fractions in soils under pine and eucalyptus forest in northwest Spain and its effect on water repellency. *Geoderma* 155: 242–248.

Carabassa, V., Domene, X., Díaz, E., and Alcañiz, J.M. (2020). Mid-term effects on ecosystem services of quarry restoration with Technosols under Mediterranean conditions: 10-year impacts on soil organic carbon and vegetation development. *Restoration Ecology* 28: 960–970. https://doi.org/10.1111/rec.13072.

Chapin, F.S., Matson, P.A., Mooney, H.A., and Vitousek, P.M. (2002). *Principles of Terrestrial Ecosystem Ecology*. New York: Springer.

Chatterjee, A., Lal, R., Shrestha, R.K., and Ussiri, D.A.N. (2009). Soil carbon pools of reclaimed minesoils under grass and forest landuses. *Land Degradation and Development* 20: 300–307.

Chenu, C. and Plante, A.F. (2006). Clay-sized organo-mineral complexes in a cultivation chronosequence: revisiting the concept of the 'primary organo-mineral complex'. *European Journal of Soil Science* 7: 596–607.

Das, R. and Maiti, S.K. (2016). Importance of carbon fractionation for the estimation of carbon sequestration in reclaimed coalmine soils – a case study from Jharia Coalfields, Jharkhand, India. *Ecological Engineering* 90: 135–140.

De Vos, B., Vandecasteele, B., Deckers, J., and Muys, B. (2005). Capability of loss-on-ignition as a predictor of total organic carbon in non-calcareous forest soils. *Communications in Soil Science and Plant Analysis* 36 (19–20): 2899–2921.

Ericsson, M. and Löf, O. (2019). Mining's contribution to national economies between 1996 and 2016. *Mineral Economics* 32 (2): 223–250.

Frouz, J., Cienciala, E., Pizl, V., and Kalcic, J. (2009). Carbon storage in post-mining forest soil, the role of tree biomass and soil bioturbation. *Biogeochemistry* 94: 111–121.

Ghani, A., Dexter, M., and Perrott, K.W. (2003). Hot-water extractable carbon in soils: a sensitive measurement for determining impacts of fertilisation, grazing and cultivation. *Soil Biology and Biochemistry* 35: 1231–1243.

Ghosh, D., Masto, R.E., and Maiti, S.K. (2020). Ameliorative effect of Lantana camara biochar on coal mine spoil and growth of maize (Zea mays). *Soil Use and Management* 36 (4): 726–739.

Hassink, J., Whitmore, A.P., and Kubát, J. (1997). Size and density fractionation of soil organic matter and the physical capacity of soils to protect organic matter. *European Journal of Agronomy* 7 (1–3): 189–199.

Heimann, M. and Reichstein, M. (2008). Terrestrial ecosystem carbon dynamics and climate feedbacks. *Nature* 451: 289.

Hoogsteen, M.J., Lantinga, E.A., Bakker, E.J. et al. (2015). Estimating soil organic carbon through loss on ignition: effects of ignition conditions and structural water loss. *European Journal of Soil Science* 66 (2): 320–328.

Huang, X., Jia, Z., Guo, J. et al. (2019). Ten-year long-term organic fertilization enhances carbon sequestration and calcium-mediated stabilization of aggregate-associated organic carbon in a reclaimed Cambisol. *Geoderma* 355: 113880.

Hurteau, M.D., North, M.P., Koch, G.W., and Hungate, B.A. (2019). Opinion: managing for disturbance stabilizes forest carbon. *Proceedings of the National Academy of Sciences* 116 (21): 10193–10195.

IEA (2020). International energy outlook. https://www.iea.org/reports/world-energy-outlook-2020 (accessed 23 April 2021).

Ingram, L.J., Schuman, G.E., Stahl, P.D., and Spackman, L.K. (2005). Microbial respiration and organic carbon indicate nutrient cycling recovery in reclaimed soils. *Soil Science Society of America Journal* 69: 1737–1745.

IPCC (2000). *Intergovernmental Panel on Climate Change 2000. Land Use, Land Use Change and Forestry*. Special Report. Cambridge, UK.: Cambridge University Press.

IUSS-WRB (2015). World reference base for soil resources. https://www.fao.org/soils-portal/soil-survey/soil-classification/world-reference-base/en/ (accessed 20 April 2021).

Jacinthe, P.A. and Lal, R. (2007). Carbon storage and minesoil properties in relation to topsoil application techniques. *Soil Science Society of America Journal 71*: 1788–1795.

Jagadamma, S. and Lal, R. (2010). Distribution of organic carbon in physical fractions of soils as affected by agricultural management. *Biology and Fertility of Soils* 46 (6): 543–554.

Jozefowska, A., Pietrzykowski, M., Wos, B. et al. (2017). The effects of tree species and substrate on carbon sequestration and chemical and biological properties in reforested post-mining soils. *Geoderma* 292: 9–16.

Juwarkar, A.A., Yadav, S.K., Thawale, P.R. et al. (2009). Developmental strategies for sustainable ecosystem on mine spoil dumps: a case of study. *Environmental Monitoring and Assessment* 157: 471–481.

Kumar, S., Maiti, S.K., and Chaudhuri, S. (2015). Soil development in 2–21 years old coalmine reclaimed spoil with trees: a case study from Sonepur-Bazari opencast project, Raniganj Coalfield, India. *Ecological Engineering* 84: 311–324.

Kumari, S. and Maiti, S.K. (2019). Reclamation of coalmine spoils with topsoil, grass, and legume: a case study from India. *Environmental Earth Sciences* 78: 429. https://doi.org/10.1007/s12665-019-8446-2.

Laird, D.A., Martens, D.A., and Kingery, W.L. (2001). Nature of clay-humic complexes in an agricultural soil: I. chemical, biochemical, and spectroscopic analyses. *Soil Science Society of America Journal* 65: 1413–1418.

Liu, X., Bai, Z., Zhou, W. et al. (2017). Changes in soil properties in the soil profile after mining and reclamation in an opencast coal mine on the Loess Plateau, China. *Ecological Engineering* 98: 228–239.

Maiti, S.K., Kumar, A., and Ahirwal, J. (2016). Bioaccumulation of metals in timber and edible fruit trees growing on reclaimed coal mine overburden dumps. *International Journal of Mining, Reclamation and Environment* 30: 231–244.

Mukhopadhyay, S. and Masto, R.E. (2016). Carbon storage in coal mine spoil by *Dalbergia sissoo* Roxb. *Geoderma* 84: 204–213.

Mukhopadhyay, S. and Masto, R.E. (2019). Effect of fly ash on carbon mineralization of biochar and organic manures added to mine spoil. *SN Applied Sciences* 1 (9): 1–10.

Muñoz-Rojas, M., Román, J.R., Roncero-Ramos, B. et al. (2018). Cyanobacteria inoculation enhances carbon sequestration in soil substrates used in dryland restoration. *Science of the Total Environment* 636: 1149–1154.

Nyirambangutse, B., Zibera, E., Uwizeye, F.K. et al. (2017). Carbon stocks and dynamics at different successional stages in an Afromontane tropical forest. *Biogeosciences* 14: 1285.

Ohsowski, B.M., Dunfield, K., Klironomos, J.N., and Hart, M.M. (2018). Plant response to biochar, compost, and mycorrhizal fungal amendments in post-mine sandpits. *Restoration Ecology* 26 (1): 63–72.

Ojeda, G., Ortiz, O., Medina, C.R. et al. (2015). Carbon sequestration in a limestone quarry mine soil amended with sewage sludge. *Soil Use and Management* 31 (2): 270–278.

Palmer, F.L., Eggins, B.R., and Coleman, H.M. (2002). The effect of operational parameters on the photocatalytic degradation of humic acid. *Journal of Photochemistry and Photobiology A: Chemistry* 148: 137–143.

Pan, Y., Birdsey, R.A., Fang, J. et al. (2011). A large and persistent carbon sink in the world's forests. *Science* 333: 988–993.

Pietrzykowski, M., Gruba, P., and Sproull, G. (2017). The effectiveness of Yellow lupine (Lupinus luteus L.) green manure cropping in sand mine cast reclamation. *Ecological Engineering* 102: 72–79.

Placek-Lapaj, A., Grobelak, A., Fijalkowski, K. et al. (2019). Post–mining soil as carbon storehouse under polish conditions. *Journal of Environmental Management* 238: 307–314.

Post, W.M. and Kwon, K.C. (2000). Soil carbon sequestration and land-use change: processes and potential. *Global Change Biology* 6: 317–327.

Qu, J.F., Hou, Y.L., Ge, M.Y. et al. (2017). Carbon dynamics of reclaimed coal mine soil under agricultural use: a chronosequence study in the Dongtan mining area, Shandong Province, China. *Sustainability* 9 (4): 629.

Rovira, P. and Vallejo, V.R. (2002). Labile and recalcitrant pools of carbon and nitrogen in organic matter decomposing at different depths in soil: an acid hydrolysis approach. *Geoderma* 107: 109–141.

Rovira, P. and Vallejo, V.R. (2007). Labile, recalcitrant, and inert organic matter in Mediterranean forest soils. *Soil Biology and Biochemistry* 39 (1): 202–215.

Rumpel, C., Kogel-Knabner, I., and Hüttl, R.F. (1999). Organic matter composition and degree of humification in lignite-rich mine soils under a chronosequence of pine. *Plant and Soil* 213: 161–168.

Silveira, M.L., Comerford, N.B., Reddy, K.R. et al. (2008). Characterization of soil organic carbon pools by acid hydrolysis. *Geoderma* 144 (1-2): 405–414.

Simón, M., García, I., Sánchez, J.A., and González, V. (2018). Atypical morphology of Technosols developed in quarry dumps restored with marble sludge: implications for carbon sequestration. *Catena* 160: 50–56. https://doi.org/10.1016/j.catena.2017.08.018.

Song, B., Niu, S., Zhang, Z. et al. (2012). Light and heavy fractions of soil organic matter in response to climate warming and increased precipitation in a temperate steppe. *PloS One* 7: 33217.

Shrestha, R.K. and Lal, R. (2010). Carbon and nitrogen pools in reclaimed land under forest and pasture ecosystems in Ohio, USA. *Geoderma* 157: 196–205.

Singh, A.N., Zeng, D.H., and Chen, F.S. (2006). Effect of young woody plantations on carbon and nutrient accretion rates in a redeveloping soil on coalmine spoil in a dry tropical environment, India. *Land Degradation & Development 17*: 13–21.

Tripathi, N., Singh, R.S., and Nathanail, C.P. (2014). Mine spoil acts as a sink of carbon dioxide in Indian dry tropical environment. *Science of the Total Environment* 468: 1162–1171.

UNFCCC (2010). United Nations Framework Convention on Climate Change. Report of the Conference of the Parties on its Sixteenth Session, Cancun (29 November to 10 December 2010). FCCC/CP/2010/7/Add.1.

Ussiri, D.A. and Lal, R. (2005). Carbon sequestration in reclaimed minesoils. *Critical Reviews in Plant Sciences* 24: 151–165.

Ussiri, D.A. and Lal, R. (2008). Method for determining coal carbon in the reclaimed minesoils contaminated with coal. *Soil Science Society of America Journal* 72: 231–237.

Ussiri, D.A., Jacinthe, P.A., and Lal, R. (2014). Methods for determination of coal carbon in reclaimed minesoils: a review. *Geoderma* 214: 155–167.

Vieira, F.C.B., Bayer, C., Zanatta, J.A. et al. (2007). Carbon management index based on physical fractionation of soil organic matter in an Acrisol under long-term no-till cropping systems. *Soil and Tillage Research* 96: 195–204.

Vinduskova, O. and Frouz, J. (2013). Soil carbon accumulation after open-cast coal and oil shale mining in Northern Hemisphere: a quantitative review. *Environmental Earth Sciences* 69: 1685–1698.

Wang, J., Wang, H., Cao, Y. et al. (2016). Effects of soil and topographic factors on vegetation restoration in opencast coal mine dumps located in a loess area. *Scientific Reports* 6: 22058.

Yan, M., Cui, F., Liu, Y. et al. (2020a). Vegetation type and plant diversity affected soil carbon accumulation in a postmining area in Shanxi Province, China. *Land Degradation & Development 31*: 181–189.

Yan, M., Fan, L., and Wang, L. (2020b). Restoration of soil carbon with different tree species in a post-mining land in eastern Loess Plateau, China. *Ecological Engineering* 158: 106025.

Yuan, Y., Zhao, Z., Zhang, P. et al. (2017). Soil organic carbon and nitrogen pools in reclaimed mine soils under forest and cropland ecosystems in the Loess Plateau, China. *Ecological Engineering* 102: 137–144.

Yuan, Y., Zhao, Z., Li, X. et al. (2018a). Characteristics of labile organic carbon fractions in reclaimed mine soils: evidence from three reclaimed forests in the Pingshuo opencast coal mine, China. *Science of the Total Environment* 613: 1196–1206.

Yuan, Y., Zhao, Z., Niu, S. et al. (2018b). Reclamation promotes the succession of the soil and vegetation in opencast coal mine: a case study from *Robinia pseudoacacia* reclaimed forests, Pingshuo mine, China. *Catena 165*: 72–79.

5

Assessing Mine Restoration Success Using Biological Soil Quality Indicators

Abstract

Increase in anthropogenic degradation of natural ecosystems is of prime concern to maintain ecological sustainability. Surface mining activities are continuously increasing to fulfil the energy demands of the growing population. At the same time, it is resulting in ecosystem disturbance and causes extensive environmental stress, loss of biodiversity, and environmental pollution. Therefore, it is vital to restore the mine degraded lands for sustainable environmental development. In this chapter, we explore the frequently used soil biological properties as an indicator parameter of the mine soil quality. Soil physicochemical characteristics were used to evaluate the status of the mine soil quality, but the biological parameters indicate pedological functions such as nutrient cycling and respond quickly to any changes in soil functions. Soil biological properties such as enzyme activity, microbial biomass carbon, soil respiration, and soil organic matter are briefly discussed. Soil biological indicators can widely contribute to a better understanding of the effect of different restoration practices and help in developing soil conservation strategies.

Keywords *Soil quality; biological properties; mine soil; restoration; indicators*

CONTENTS

Eco-Restoration of Mine Land, First Edition. Vimal Chandra Pandey, Jitendra Ahirwal, Roopali Roychowdhury, and Ritu Chaturvedi.

5.1 Introduction

Surface mining is an essential economic activity carried out in many parts of the world to excavate natural resources from the Earth's crust. The extraction can be carried out either as open-cut mining and/or underground mining. Mining methods depend on the geology, types of minerals, and economy of the states and, thus, can vary in different countries largely. Surface mining methods are widely used because they are economical and highly efficient compared to underground mining. Surface mining is mainly undertaken in those areas where minerals are deposited at shallow depth. However, surface mining is associated with ecological and social consequences such as land degradation, deforestation, acid mine drainage, groundwater contamination, soil erosion, metal toxicity, and relocation of local population (Shukla and Lal 2005; Maiti 2013; Feng et al. 2019). In opencast mining, initially, the land is cleared by vegetation removal, and topsoil and subsoil (upper 30–100 cm) layers are scrapped and used either concurrently or stored in a topsoil dump for future reuse (Maiti 2013). Removal of entire vegetation, soil cover, and overlying rocks to mine the land results in permanent loss of soil fertility, carbon sink, land use change, and biodiversity loss that damages and pollutes the environment (Ussiri et al. 2006; Ahirwal et al. 2021). The release of air pollutants, greenhouse gasses, and trace elements can also be an important environmental concern associated with surface mining activities.

Historically mine reclamation aimed to reinstate ecological conditions by plantation. However, to comply with environmental laws and produce more ecosystem benefits, the paradigm shift was observed to restore site conditions before mining or can mimic the natural ecosystem. Restoration to the pre-disturbed condition depends on the types of practices adopted to restore land. One of the major concerns of mine restoration is to restore mine overburden dumps (mine spoil) which is large in volume and occupy nearby area during dumping (Figure 5.1). The properties of mine spoils are dependent on the physical and geochemical properties of the parent material. Mine spoils consist of coarse rock fragments and low moisture content and may contain trace metals depending on the geological formation of the rock (Barré et al. 2017). Once the mine spoils are physically reclaimed and used as plant growth medium, they are known as "mine soils" (Sencindiver and Ammons 2000).

Figure 5.1 Land degradation and deforestation due to surface mining activities; (a) typical view of the surface mining of coal, (b) coal mine spoil dump in the vicinity of coal mining, (c) iron ore mining showing benching and iron ore seam, and (d) iron ore dumpsite in natural forest.

Restoration of mine spoil to mine soil takes time. Therefore, periodic assessment of mine soil quality through physical, chemical, and biological characterization is essential to determine the effect of applied amendments and natural establishment of an ecosystem. Soil physical properties such as texture, bulk density, moisture, pore space, and water holding capacity develop slowly because it depends on the gradual changes in soil material, organic matter supply, and biological functioning of the soils. Soil's physical, chemical, and biological properties are interconnected, and changes in a single parameter led to changes in other soil properties. Similarly, improvement in soil chemical properties such as pH, electrical conductivity, base cations, trace elements, and nutrients largely depend on the vegetation growth and soil amendments. Both physical and chemical properties rely on biological properties (Maiti 2013). Thus, improvement in biological properties and the adoption of sustainable practices to enhance soil biological characteristics are vital to mine land restoration. In the chapter, we discuss major soil biological properties that regulate biogeochemical cycles and act as indicator parameters to determine mine soil quality and restoration success.

5.2 Revegetation of Mine Lands

Revegetation of mined land involves the reconstruction of disturbed ecosystems or creating site conditions that support the establishment of vegetation and are comparable to the natural ecosystems or before land disturbance. Surface mining cleared large patches of forest that reduce biodiversity, deteriorate soil quality, and influenced water and nutrient cycling. These disturbed lands can be restored by complex technical and biological restoration practices (Ahirwal et al. 2016; Maiti and Ahirwal 2018). This encompasses reconstruction of mine land, landscape design to stable land, soil amendments to provide a suitable growth medium for vegetation development, fertilization, and plantation of selected tree species to create a self-driven ecosystem. Although revegetation is essential, mine wastes are very hostile environments for plant growth due to the abiotic constraints such as lack of soil structure, presence of toxic metals, unstable topography, and presence of several growth-limiting factors. Revegetation plays a crucial role in the restoration of mine spoil because it improves the soil characteristics, provides a substrate for microbial decomposition, and supplies nutrients into the soils. Nevertheless, it is the oldest and most widely used technique to restore mine soil fertility and the ecological integrity of degraded lands.

Revegetation of mine wasteland involves both ameliorative and adaptive strategies that help in plant establishment and encourage subsequent vegetation development (Tordoff et al. 2000). Several native and fast-growing tree species were used for the mine land revegetation (Table 5.1).

The tree species can vary widely across the globe as it depends on the type of mine, climatic conditions, and interests of the stakeholders. For example, five acid-tolerant plants: *Cynodon dactylon*, *Pennisetum purpureum*, *Neyraudia reynaudiana*, *Panicum repens*, and the non-native *Eucalyptus robusta* along with soil amendments were introduced at a metal sulfide-contaminated acidic mine waste in China (Yang et al. 2016). The species mixture contains native grass species and invasive tree species that were established successfully in acidic conditions and increased pH, organic matter, and nutrient stock only after six months. Similarly, *Pinus sylvestris* plantation on surface mine spoils of lignite, sulfur, sand, and hard coal mine spoils improve soil quality and sequester a significant amount of carbon in soils and plant biomass (Pietrzykowski and Daniels, 2014). Development of an eco-restoration strategy on kaolin, quartz sand, and iron mining degraded lands in Romania over 40 years through different plant cover (natural grassland, pasture cover, black locust, Norway spruce, and Scots pine) reported

Table 5.1 Tree species commonly used for revegetation of post-mining lands globally.

Species	Mine type	Country
Lotus corniculatus L., *Acer rubrum* L., *Rhus glabra* L., *Trifolium repens* L., *Robinia pseudoacacia* L.	Coal	USA
Euphorbia cyparissias L., *Indigofera gerardiana* (Wall.) Baker, *Desmodium triflorum* L., *Euphorbia ischeriana* Steud.	Coal	Russia
Pinus armandii Franch, Populus tomentosa Carriere, Robinia pseudoacacia;	Coal	China
Acer spp., *Fraxinus excelsior, Larix decidua*	Coal	Czech Republic
Anadenanthera peregrina, Eucalyptus spp., *Bracatinga scabrella* Benth,	Bauxite	Brazil
Pinus pinea, Nerium oleander	Pyrite	Spain
Alnus alnobetula ssp., *Alnus incana* ssp.	Gold	Canada
Picea glauca Voss, *Populus tremuloides* Michx.	Oil sand	Canada
Alnus incana, A. viridis, A. glutinosa	Sand	Poland
Pinus sylvestris L., *Larix decidua* Mill., *Quercus robus* L., *Alnus* ssp.	Lignite and Sulfur	Poland, Central Europe
Artemisia sieberi, Salsola yazdiana, Halophytes types	Iron	Iran
Alnus nepalensis, Cupressus torulosa	Phosphate	China

improvement in soil quality where pasture cover significantly increased microbial activity, soil organic carbon (SOC), and nitrogen (N) and phosphorus (P) concentration (Buta et al. 2019).

Fast-growing invasive tree species were often used in mine revegetation. Currently, the focus has been shifted to planting multipurpose tree (MPT) species, which also increases the economy of the local peoples. For example, several tree species were recommended for mine restoration in India (Table 5.2). It is vital to select particular species or group of species demonstrating a wide range of adaptive characteristics such as drought tolerant, heat resistant, and pH tolerant. Mine spoil often exhibits adverse physicochemical characteristics and lacks soil structure that inhibits plant growth and microbial activity. Most of the plant species used for afforestation of mined land belongs to the Fabaceae family because they are fast-growing, are nitrogen-fixing plants, accumulate a significant amount of carbon in biomass and soils, and tolerate mine spoil conditions (Mukhopadhyay and Masto 2016; Ahirwal et al. 2020). Bio-reclamation of coal mine waste by planting *Dalbergia sissoo, Eucalyptus*, and *Bamboo* species in India

Table 5.2 Tree species commonly used for revegetation of post-mining lands in India.

Species	Family
Acacia auriculiformis A. Cunn.	Fabaceae
Acacia catechu (L. f.) Willd.	Fabaceae
Ailanthus excelsa Robx.	Simaroubaceae
Alstonia scholaris L.	Apocynaceae
Albizia lebbeck L. (Benth)	Fabaceae
Albizia procera Roxb. (Benth)	Fabaceae
Azadirachta indica A. Juss.	Meliaceae
Butea monosperma (Lam.)	Fabaceae
Bombax Ceiba L.	Malvaceae
Cassia fistula L.	Fabaceae
Cassia siamea (Lam.)	Fabaceae
Dalbergia sissoo Roxb.	Fabaceae
Dendrocalamus spp.	Poaceae
Eucalyptus spp.	Myrtaceae
Ficus racemosa L.	Moraceae
Gmelina arborea Roxb.	Lamiaceae
Heterophragma adenophyllum Seem.	Bignoniaceae
Holoptelea integrifolia (Roxb.) Planch	Ulmaceae
Leucaena leucocephala (Lam.) de Wit	Fabaceae
Mangifera indica L.	Myrtaceae
Madhuca indica J.F. Gemel	Sapotaceae
Melia azadarech L.	Meliaceae
Millettia pinnata (L.) Panigrahi	Fabaceae
Pletophorum pterocarpum (DC.) K. Heyne	Fabaceae
Prosopis juliflora (Sw.) DC.	Fabaceae
Psidium gujava L.	Myrtaceae
Syzygium cumini (L.) Skeels	Myrtaceae
Tectona grandis L.f.	Lamiaceae
Terminalia arjuna (Roxb.) W. & A.	Combretaceae
Trema orientalis L. Blume	Cannabaceae

improved soil pH, moisture, bulk density, and nutrients content and demonstrated a higher accumulation of toxic metals in different planted species (Maiti 2007). Planting these species have advantages over others because they not only improve soil quality but also reduce metal toxicity. Similarly, a combination of fruits (*Mangifera indica, Anacardium occidentale, Psidium guajava, Artocarpus heterophyllus,* and *Syzygium cumini*) and timber trees (*Acacia mangium, Eucalyptus* spp., *Tectona grandis*, and *Gravellia robusta*) was used for the reclamation of surface mine lands that improves soil quality and reduce metal toxicity (Maiti et al. 2016). Coal-mined land was afforested with xerophytic plant, *Prosopis juliflora*, that successfully established on dry and nutrient deficit mine spoil and accumulated a significant amount of carbon and nutrients stock after eight years of afforestation (Ahirwal et al. 2017). Therefore, it is vital to select specific plant species that provide multitudinous (ecological, economic, and social) services in mine restoration.

Continuous nutrient supply and development of soil structure are essential for vegetation growth. Tree species took one to two years to ameliorate rhizosphere soil and establish themselves in a mining environment. It needs fertilizers, soil amendments, and continuous care after planting which increases the cost and time of restoration and N emissions. However, the growth of legume species along with other woody vegetation increases biomass yield comparable to mineral N fertilizer reducing N emissions and reducing the chances of weed invasion owing to thick vegetation cover (Yang et al. 2019; Kocira et al. 2020). Therefore, it is generally recommended to use a mixture of grass–legume species at the initial stage of revegetation because it helps in biomass accumulation, soil particle binding, and mine waste stabilization (Maiti 2013). Some efficient grass and legume species used to restore the mine degraded land in Indian conditions are *Crotalaria spectabilis, Cymbopogon citratus, Stylosanthes humilis, Hibiscus sabdariffa, Pennisetum pedicellatum, Sesbania sesban, Stylosanthes hamata,* and *Vetiveria zizanioides*. A typical view of the grass–legume mixture growing on mine overburden dump is shown in Figure 5.2. In addition, short-living and annual grasses supply nutrients through biomass production and its subsequent decomposition that reduce cost of fertilization.

In many countries, the use of understory vegetation (grass, shrub, and herbs) has been proven to be effective in the initial restoration of mined lands (Yang et al. 2016; Kumari and Maiti 2019). Based on the site conditions, understory species may colonize at the post-mining sites.

The gradual establishment of understory vegetation contributes to increasing biodiversity and ecosystem functioning and may be used as an ecological indicator of restoration success (Alday et al. 2014; Dhar et al. 2020). A review of the

Figure 5.2 A distance view of the grass–legume mixture growing on mine overburden dump.

Table 5.3 Types of grass and shrub species commonly used for revegetation of post-mining lands.

Species	Mine type	Country
Cenchrus ciliaris L	Coal	Colombia
Lolium multifloram, Panicum virginatum, Lolium multiflorum, Lolium perenne	Coal, gold, and silver	USA, Canada
Medicago sativa, Caragana microphylla	Coal	China
Avena sativa, Phleum pretense	Asbestos	Canada
Alopecurus pratensis	Iron and uranium	Poland
Cannabis sativa L.	Lignite	Poland
Miscanthus sinensis	Lead-zinc	China
Pennisetum pedicellatum	Coal	India
Cymbopogon citratus, Chrysopogon zizanioides, Cynodon dactylon	Chromite-asbestos	India

literature shows that numerous native and invasive understory species have been found successfully grown on different mined lands (Table 5.3). Application of legume (*S. hamata*) and grass (*Cenchrus ciliaris*) species in an early stage of reclamation show a significant impact on mine soil fertility and higher organic matter and N accumulation under legumes (Kumari and Maiti 2019).

5.3 Reclaimed Mine Soil Quality Indicators

Criteria for judging restoration trajectories of disturbance and restoration of land require specific parameters. Currently, the importance of soil quality assessment using indicator parameters has largely been adopted to determine the effects of different management practices on soil quality. Indicators of complex terrestrial ecosystems which are determined by processes are of special importance, although they can be time consuming and difficult to measure directly. Indicators of soil quality should be easy to recognize, inexpensive, and simple to evaluate by the land managers. Measurement of these indicator parameters should also indicates why the soil is not functioning as desired. In general, some of the soil's physical, chemical, and biological properties were considered indicator parameters (Figure 5.3).

Soil quality indicators can differ among the ecosystems and management practices. For example, the effect of anthropogenic pressure on the soil quality of a national park in Southern Italy showed that only 4 parameters out of 25; carbon content, total nickel concentration, qCO_2, and percentage effect of root elongation were sufficient to evaluate soil quality (Memoli et al. 2019). Quantification of changes in soil quality due to Chinese fir plantations on forest land of southern China reported total phosphorus, labile carbon pool, and pH as key soil quality indicators for temporal soil quality monitoring under forest plantations (Zhijun et al. 2018). The study of soil quality indicators of the land use change proposed that bulk density, plant available water, wide coarse pores, air conductivity, SOC,

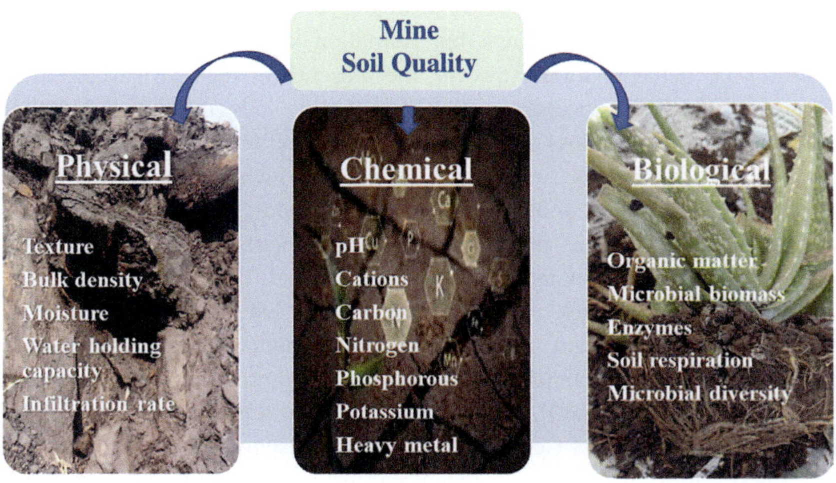

Figure 5.3 Soil parameters used to evaluate mine soil quality.

Al, P, pH, and base sum can be used to evaluate changes in volcanic soils (Valle and Carrasco 2018). As there is no unanimously used indicator parameter to determine the soil quality, there is a need for standardization of the indicators that can be frequently used to assess the soil quality. A rapid, reliable method for the analysis of mine soil quality is needed to make soil protection policies related to fertility, erosion control, and protect soil biodiversity.

Soil biological property or group of properties may be able to reflect the status of the ecosystem processes sensitive to environmental changes and variation in the management practices. Biological approaches, particularly soil microbial activity in terms of biomass, respiration, enzyme activity, and diversity, are widely used to evaluate soil quality and restoration success (Table 5.4). Further attention is required by the restoration ecologist to develop ecological indicators for mine restoration success.

5.3.1 Soil Organic Matter

Soil is the largest sink of terrestrial carbon and the third-largest global carbon reservoir. Soil organic matter (SOM) is an important component of the soil as it influences various soil properties and is essential for nutrient availability to plants. The carbon content of SOM ranged from 40 to 60%; therefore, it plays a vital role in the global carbon cycle (Palmer et al. 2002). The estimation of SOM has often been used to determine the impact of anthropogenic or natural disturbance on soil quality. Surface mine land has been subjected to various alterations such as a change in soil structure and fertility loss, which altogether affect the mine soil functions. The loss of SOM due to mining activities and their accretion during the reclamation process is of great importance in biogeochemical cycling and fertility of the mine soil. Studies have shown that estimation of SOC is generally used as a proxy for SOM determination; however, the presence of the coal carbon often overestimates the SOC concentration in the coal mine soils (Ingram et al. 2005). There has been a massive loss of SOC during the topsoil removal and stockpiling during the surface mining (Graham and Haynes 2004) which may be due to the mineralization of the SOM and lack of fresh carbon input (Banning et al. 2008).

Several strategies such as plantation and amendments with recycled waste have been developed to increase SOC in mined land, but most of the time, adverse mine soil properties retard the plant growth. Estimation of the SOC in manganese mine reclaimed with the integrated biotechnological approach showed that SOC in 0–15 cm profile depth increased approximately 7 times over 20 years (Juwarkar et al. 2010). While investigating the organic matter dynamics in lignite-rich mine soils of Germany under a chronosequence of pine, Rumpel et al. (1999) found an increase in recent carbon accumulation from 11 years ($1284\,g\,m^{-2}$) to 32 years ($5018\,g\,m^{-2}$) in the top 5 cm profile, which is much higher

Table 5.4 Soil biological parameters used as an indicator to assess reclaimed mine soil quality.

Indicator parameters	Objective of analysis	Authors
Enzyme activity	Soil rehabilitation process	Ciarkowska et al. (2014)
	Substrate quality in revegetation	Finkenbein et al. (2013)
	Potential biomarkers of metal stress	Hinojosa et al. (2004) and Wang et al. (2010)
	Soil restoration after revegetation	Izquierdo et al. (2005)
	Technosols quality assessment	Ahirwal et al. (2018)
	Bioindicator of coal-contaminated soil	Masto et al. (2011)
Microbial diversity	Functionality of restored soil	Muñoz-Rojas et al. (2016)
	Indicators of reclamation progress	Mummey et al. (2002)
	Soil health indicators	Li et al. (2015a)
Microbial biomass carbon (MBC)	Soil quality in metal mine	Zhang et al. (2010)
Microbial metabolic quotient (qCO_2)	Stresses in mine soils	Liao and Xie (2007)
Soil respiration	Indicator of soil microbial activity	Goupil and Nkongolo (2014)
	Biological activity of soils subject to reclamation	Uzarowicz et al. (2020)
	Evaluation of mine reclamation success	Mukhopadhyay et al. (2014)
	Reclaimed mine soil health	Haney et al. (2008)
Soil organic carbon/matter	Mine sand dunes rehabilitation	Graham and Haynes (2004)
	Carbon storage in coal mine soil	Mukhopadhyay and Masto (2016)

than the natural forest soil (3827 g m^{-2}). Over 21 years, organic matter accumulation in the reclaimed topsoil (0–7.5 cm) was found higher than that in the undisturbed soil (Malik and Scullion 1998). With time, the development of vegetation cover supplies the fresh carbon to mine soil that is later decomposed by the microorganisms and increases the mine soil fertility. A significant increase in SOC from 1.4–6.6 to 10–112 g kg^{-1} in the reclaimed copper mine soils of Spain was due to the tree vegetation and waste management (Asensio et al. 2014). Mine soils often showed a low SOC concentration that promises the high SOC sequestration potential with a time that can be a sensitive indicator of the mine soil quality. It is recommended that the minimum concentration of 0.52% SOC or 0.89% of SOM is essential to maintain the nutrient cycle in the RMS (Ingram et al. 2005).

5.3.2 Soil Microbial Biomass

Microbial biomass is a part of living SOM that reflects the total mass of microorganisms (Srivastava and Singh 1989). It is accountable for organic matter transformation and nutrient cycling which in turn influence and impact plant growth and soil health. But mining activities significantly reduce microbial biomass, richness, and diversity in soils (Quadros et al. 2016). The establishment and survival of plant communities in mined land restoration are highly dependent on microbial activities. Microorganisms have the potential to increase the nutrient content through mineralization, although the absence of nutrient content retards the growth of microorganisms (Singh et al. 1989). Microbial biomass carbon (MBC) and organic matter content act as nutrients to plants and play a vital role during the early stages of succession. Plants provide root exudates and organic matter that encourage the development of biological communities in the mine soils, which in turn it provide nutrients for plant growth (Srivastava and Singh 1989). Studies have reported the effect of tree plantations, soil amendment, and vegetation growth on soil microbial activities and increase in soil microbial biomass (Table 5.5).

Table 5.5 Status of soil microbial biomass carbon in the coal mine dumps.

Study area	Age (years)	Climate	MBC (mg C kg^{-1})	Authors
Czech Republic	8–40	Temperate	200–600	Šourková et al. (2005)
Greece	1	Subhumid Mediterranean	70–120	Monokrousos et al. (2014)
Poland	—	Temperate	258–442	Chodak and Niklińska (2010a)
India	2–17	Dry tropical	92–388	Mukhopadhyay et al. (2014)
India	10	Dry tropical	50–600	Sinha et al. (2009)
India	2–21	Tropical humid	98–368	Kumar et al. (2015)
India	4–6	Tropical monsoon	124–363	Singh et al. (2004)
India	5–20	Tropical monsoon	209–867	Srivastava and Singh (1989)

Physical and chemical factors such as pH, moisture, and nutrient content influence the microbial activity in mine soil. For example, an elevated concentration of heavy metals severely affects the growth of microorganisms by reducing the biochemical activity and microbial diversity (Piotrowska-Seget et al. 2005). In contrast, agronomic practices like mulching increased MBC by 18% in a revegetated coal mine spoil compared to control soils (Shrestha et al. 2019). Similarly, a chronosequence study of mine revegetation in dry tropical climate reported a 14-fold increase in MBC concentration after 3–15 years of revegetation (Ahirwal et al. 2018). MBC status in different mine soils can vary among the mine type and reclamation age. Moreover, it is largely depending on the soil amendment, plant species, and availability of soil substrate.

5.3.3 Soil Enzymes Activity

Soil enzyme activity is the sensitive indicator of anthropogenic disturbance and an essential criterion to assess the health of the restored ecosystem (Dick 1997). It is an expression of the soil microbial activities that instigate the biogeochemical transformation of the soil, organics, and mine spoil components. The enzyme level in mine soil may vary due to the presence of substrate and microbial activities. Since it controls the supply of nutrients to plants and microbial growth, it may be used to measure the soil's biological potential. Most of the energy resources are present in the bound form in the environment that cannot be directly used by the soil microbes. Soil microbes release extracellular enzymes that metabolize organic matter and release nutrients in a simple form that can be directly utilized by the microbial cell to modify the microenvironment (Allison and Vitousek 2005). The function of mine soil enzymes mainly includes their role in soil development (Zhan and Sun 2014) and mineralization of organic matter to release C and nutrients for plant growth (Baldrian et al. 2008). Enzymes are mainly released by the soil microorganisms in response to the substrate, and therefore may serve as an indirect measure of soil microbial biomass (Sinha et al. 2009).

Several enzyme activities depend on environmental factors. Internal and external factors such as climatic conditions and soil properties notably alter the enzyme activity. Dehydrogenase, phosphatase, β-glucosidase, urease, and arylsulfatase activities are generally determined to assess C, N, P, and S cycles, respectively. Studies on the effects of arbuscular mycorrhizal fungi (AMF) along with the coal gauge, fly ash, and sludge on the enzyme's activity of the reclaimed coal mine soils of China reported an increase in saccharase (44–53%), dehydrogenase (23–43%), urease (36–54%), and phosphatase (28–49%) enzyme activities after a year of soil amendments (Qian et al. 2012).

Effects of vegetation development after 18 years on soil enzyme activity in reclaimed coal mine of Northwest Loess Plateau of China show an increase in enzyme activities viz. invertase (151–455 mg kg^{-1} h), dehydrogenase (122–172 mg kg^{-1} h), polyphenol oxidase (30.4–44.1 mg kg^{-1} h), and decrease in urease (0.17–0.14 mg kg^{-1} h) after 18 years of plantation. Similarly, enzyme activity was increased under mixed forests viz. invertase (844 mg kg^{-1} h), dehydrogenase (206 mg kg^{-1} h), polyphenol oxidase (78.6 mg kg^{-1} h), and urease (0.68 mg kg^{-1} h) compared to control site (Li et al. 2015b). Land degradation decreases enzyme activities after zinc and lead mining in southern Poland and partially recovered up to 31% (urease) and 34% (invertase) after 40 years of flora succession (Ciarkowska et al. 2014). Soil enzyme activities are strongly related to the chemical, physical, and biological properties of the mine soil. Environmental factors such as pH, temperature, moisture, resource availability, nutrient content, and organic matter; xenobiotics like pesticides and fertilizers; heavy metals; and soil ecology may affect enzyme activity. Various soil amendments (power station ash [PSA], charred rice straw [CRS], and fine material [FM]) were investigated in field trials to determine their effects on the coal mine soil fertility via changes in enzyme activity in Vietnam (Finkenbein et al. 2013). After four months of treatments, different enzyme activities showed divergent responses to each amendment. β-glucosidase, arylsulfatase, and FDA hydrolysis increased in all the treatments, while acid phosphatase, alkaline phosphatase, catalase, and arginine ammonification decreased in all treatments. Moreover, urease activity increased under PSA and decreased in CRS and FM treatment.

Miscanthus biochar in combination with lime and N fertilizer was evaluated for its ability to reclaim acidic mine spoils in the United States (Novak et al. 2018). Compared to the control site, soil *N*-acetyl-β-D-glucosaminidase enzyme activity was doubled under 1% biochar + N-fertilizer and β-glucosidase activity was increased by 10-fold under 5% biochar + lime + N-fertilizer. Reclamation of lignite mine areas in Greece with *Triticum aestivum* in chronosequence showed an increase in β-glucosidase from 0.34 to 1.21 μmol PNP g^{-1} h^{-1} after 0–15 years but decreased to 1.14 μmol PNP g^{-1} h^{-1} after 20 years. Similarly, alkaline phosphatase increased from 109 to 291 mg PNP g^{-1} h^{-1} after 0–25 years. In contrast, urease activity decreased with the age of reclamation and was found lowest after 15 years (0.47 μmol NH_4 g^{-1} h^{-1}). The study shows that after 15 years of reclamation, soil enzymatic activities converged to values recorded in the controls field (Mastrogianni et al. 2014). In a controlled pot experiment, tree (*Morus alba* L. and *Broussonetia papyrifera* L.)–herb (*Pteris vittata* L. and *Arundo donax* L.) co-planting was tested to remediate nonferrous metal(loid)-contaminated soil in China (Zeng et al. 2019). Authors reported an increase in dehydrogenase (3.48–3.81 μg g^{-1} day^{-1}), urease (57.2–85.9 μg g^{-1} day^{-1}), sucrase (0.61–4.07 μg g^{-1} day^{-1}), and acid phosphatase (8.97–191 μg g^{-1} day^{-1}) activity compared to control site that indicates the positive impact of co-planting on microbial activity.

Effects of biochar and fresh compost as individual treatment and combined effects were investigated on the enzyme activities on metal-contaminated mine soils that show enzyme activities were almost completely inhibited by biochar and activated by compost (Tang et al. 2020). Authors also confirmed that electrical conductivity and available potassium were the most important factors affecting enzyme activities, and metal availability (As, Cu, Cd, and Zn) has a significant impact on dehydrogenase, catalase, and urease activities. Soil enzyme activities were studied in soils from a degraded mining site, and two areas revegetated with *Casuarina equisetifolia* and *A. occidentale* under subtropical conditions and suggested that enzyme activity is a very sensitive indicator of the improvement in soil quality after revegetation (Izquierdo et al. 2005). Enzyme activities viz. dehydrogenase, urease, protease, phosphatase, and β-glucosidase recovered up to 33, 25, 1.5, 127, and 7.5% under *C. equisetifolia* and 36, 22, 1, 77, and 2% under *A. occidentale*, respectively. However, after 6 years of revegetation, enzyme activities were not recovered up to soils under natural vegetation but significantly increased after revegetation. Response of microbial characteristics to heavy metal pollution in mining soils under different plantations shows decreases of sucrase (1.4–8.1 g Suc kg dry soil^{-1}), urease (0.09–0.36 g NH_3 kg dry soil^{-1}), and acid phosphatase (5.3–28.8 mg PNFF kg dry soil^{-1}) enzyme activities compared to control site in the central Tibet (Zhang et al. 2010). Effect of soil texture and the planted tree species on microbial properties of mine soils reclaimed for forestry were reported higher importance of texture of the soil substrate compared to plantation types. Authors also reported higher dehydrogenase (28%), acid phosphomonoesterase (37%), and urease (56%) enzyme activities under loamy sand compared to sandy soils of the lignite mining in Central Poland (Chodak and Niklińska 2010b).

5.3.4 Mycorrhizal Fungi

Biological reclamation of spoil materials aimed to restore microbial communities and continue the nutrient cycle. Mycorrhizal fungi play a vital role in the biological interaction among plants and microorganisms. They are heterotrophs and depend on living and dead organic matter to get carbon and energy. It is reported that they can tolerate a wide range of environmental conditions from low oxygen to fewer nutrients. It is a symbiotic relationship between the fungi and host plant, where fungi provide nutrients and water to the plant and in turn, they get carbohydrates from plants. Fungi easily absorb elements such as N and P and make them available to plants. They can also assist plants in the acquisition of different nutrients present in low quantity and immobile. AMF is essential for plant nutrient acquisition and regulating nutrient cycling in the mining area (Xiao et al. 2019). A systematic review of mycorrhizal presence in mining-affected soils reported that numerous AFM species exist in mining sites and >80% of plant species are colonized by AMF (Wang 2017). The author also opined that AMF improve mineral acquisition, contribute to soil structure and function, and increase tolerance of host plant that together contribute to ecological restoration of mined lands. Thus, inoculation of AMF in mine soils can be a cost-effective method to improve soil physicochemical and biological functions and reduce the cost of chemical fertilization (Agus et al. 2018; Song et al. 2020).

AMF has a direct and indirect effect on the mine soil quality via an effect on abiotic components, such as mineral composition, pH, and nutrients, and interaction with biotic components like roots. In highly contaminated soils, this relation can protect plants from a higher accumulation of toxic metals. For example, a study on the impact of mycorrhizal colonization in maize growing in mining-affected soils increases dehydrogenase activity and chlorophyll by 40 and 10%, respectively, than the control soil (Dhawi et al. 2015). Likewise, the effect of AMF on the RMS reported an 80% increase in SOC content and a 30% increase in dehydrogenase activity (Qian et al. 2012). Another study in the coal mining region revealed that drought stress increased mycorrhizal colonization and decreased plant dry weights, nutrient contents, water use efficiency, and rehydration rate (Zhao et al. 2015). A combined effect of shrub revegetation and AMF on the coal mine soil enzyme activities were investigated in Northern China and reported urease activity in AMF soils ranged from 0.43 to 1.24 $mg\,g^{-1}\,day^{-1}$ compared to control soils (0.21–1.29 $mg\,g^{-1}\,day^{-1}$) and alkaline phosphatase from 0.20–0.98 $mg\,g^{-1}\,day^{-1}$ compared to control site (0.14–0.58 $mg\,g^{-1}\,day^{-1}$) (Xiao et al. 2019). Reestablishment of AMF after 14 years of inoculation in semi-arid surface mine soils reported a 1% increase in SOM and improvement in soil structure and a decrease in soluble salts (Frost et al. 2001). Thus, from the above observation, it can be concluded that colonization of mycorrhizal fungi helps improve mine soil characteristics in terms of physical, chemical, and biological properties.

5.3.5 Soil Respiration

Soil respiration is a major part of global carbon fluxes. Continuous increases in atmospheric CO_2 concentration can be offset through soil-based carbon sequestration. Soil can either act as a sink of carbon or a source of CO_2, and it depends on the input via plants and losses through soil respiration. Soil respiration is a sum of autotrophic and heterotrophic respiration through roots, microbes, and other soil fauna. It can vary temporally, spatially, and vertically in the soils. Mine soils in particular have lesser plant growth and adverse soil characteristics which together affect the soil respiration rate. Soil CO_2 flux measurements can help evaluate land reclamation status following surface mining. There is a limited number of studies around the world that reported soil respiration rate in RMS compared to a forest and agricultural land (Table 5.6).

Table 5.6 Soil respiration rate in different reclaimed mine soils.

Mine type	Country	Reclamation age (year)	Soil respiration ($\mu mol\ CO_2\ m^{-2}\ s^{-1}$)	Authors
Coal	Czech Republic	10	1.95	Bujalský et al. (2014)
		18	2.50	
		33	3.38	
		35	3.10	
		55	1.83	
Coal	India	3	0.98	Ahirwal et al. (2018)
		7	1.69	
		10	2.71	
		15	3.19	
Coal	USA	7	2.98	Mathiba et al. (2014)
		18	2.63	
Coal	China	2	1.80	Li et al. (2019))
Limestone	Spain	2	2.97	Soria et al. (2021)
Bauxite	Brazil	3.5	1.26	Valente et al. (2021)

Soil respiration rate in mine soils is often recorded low because of lesser microbial growth and root density. However, it can vary largely among the sites, soil carbon pool, soil amendments, and vegetation types. For example, a study from afforested mine land in Czechia reported that the relationship between CO_2 flux and CO_2 concentration is positive when root biomass, soil CO_2 concentration, and flux are high, but the relationship tends to be negative when microbial biomass and microbial respiration are high but soil porosity is low (Frouz and Bujalský 2018). A pot scale study shows that amendment with biochar reduces soil CO_2 flux with an increase in biochar concentration from 2% ($2.85\ \mu mol\ CO_2\ m^{-2}\ s^{-1}$) to 3% ($2.60\ \mu mol\ CO_2\ m^{-2}\ s^{-1}$) maybe because of higher recalcitrant carbon in biochar that reduces microbial respiration rate (Ghosh and Maiti 2021). A field study from unreclaimed and reclaimed mined land from India reported an increase in CO_2 flux after five years of revegetation ($2.37\ \mu mol\ CO_2\ m^{-2}\ s^{-1}$) compared to unreclaimed ($0.21\ \mu mol\ CO_2\ m^{-2}\ s^{-1}$) mine dump (Ahirwal and Maiti 2018).

5.4 Development and Use of Soil Quality Index

Soil quality is defined as *the capacity of a specific kind of soil to function, within natural or managed ecosystem boundaries, to sustain plant and animal productivity, and support human health and habitation* (Karlen et al. 1997). Mine soils are heterogeneous and may reflect wider soil characteristics that are different from the original/native soils. Therefore, assessment of mine soil quality is vital to determine the status of the soil functions and to predict the path of restoration of mine soil quality. Development of a soil quality index (SQI) to assess soil quality in diverse land use such as forest, agriculture, and managed ecosystems is widely practiced and several SQIs were developed in the recent past.

Soil quality is the qualitative analysis of physicochemical and biological properties that reflect the potential of soil to perform functions. Initially, several soil properties were taken into account to study soil quality. This type of assessment takes time and labor to determine a particular set of soil characteristics. Thus, the need for a specific set of soil quality indicator parameters has been arising. Statistical methods like principal component analysis (PCA) and other multivariate analyses help develop a SQI that comprises the most important soil quality parameters indicating soil quality. The development of the SQI integrates the mathematical tools and goals of soil productivity; therefore, it is a tedious task to put forward both objectives together. Development of SQI using PCA analysis involves the following steps (Figure 5.4):

i) Screening of soil properties to select the smallest group of the numeric dataset to assess the soil quality. All the analyzed soil properties are used in PCA analysis, and the results of PCA provide a specific set of soil properties having higher loading. Soil properties showing >5% variance in the dataset can be used to produce a minimum dataset.

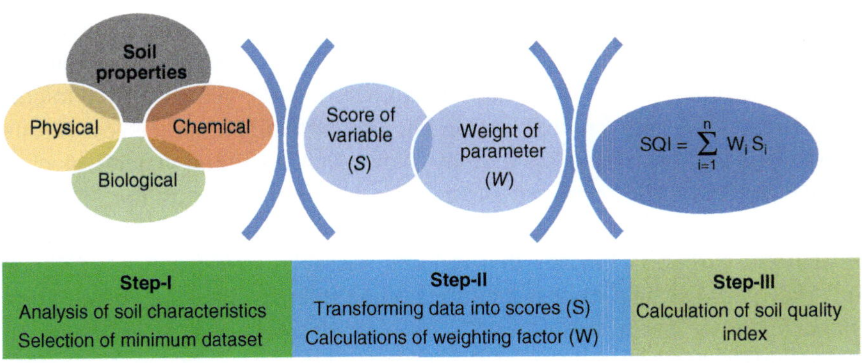

Figure 5.4 Major steps of developing a soil quality index (SQI).

ii) Transformation of minimum dataset into a numeric score can provide the score in various units into numbers called indicators. The score of an individual soil parameter can be calculated using an equation based on a sigmoidal curve with an asymptote ranging from 0 to 1:

$$S = \frac{a}{\left[1 + \left(\frac{x}{x_0}\right)^b\right]}$$

where x is the value of an individual soil property, a is the maximum score (=1) of the soil property, x_0 is the mean value of each soil property, and b is the slope of the equation.

Parameters from the MDS were ranked as good or bad in terms of their value for soil functioning. The slope was considered −2.5 for "more is better" and +2.5 for the "less is better" to obtain a sigmoidal curve (Mukhopadhyay et al. 2016).

iii) Calculation of SQI: SQI was calculated using scoring values and the weighting factor of each PC by the following weighted addition equation (Andrews and Carroll 2001):

$$\text{SQI} = \sum_{i=1}^{n} W_i S_i$$

where W_i is the weighting factor of the soil quality indices and S_i is the non-linear score.

The SQI can be used to evaluate soil quality in a particular ecosystem, the status of reclamation, screening of efficient plant species, and the impact of land management practices on soil fertility. Evaluation of reclaimed mine soil quality using PCA analysis shows that coarse fraction, soil pH, SOC, dehydrogenase activity (DHA), Ca, EC, P, and S can indicate mine soil quality under *D. sissoo* plantation (Mukhopadhyay et al. 2016). Similarly, screening of plant species for reclamation of coal mine spoil recommend *Cassia seamea* and *D. sissoo* based on their rhizosphere soil quality (Mukhopadhyay et al. 2013). A study used PCA analysis to derive rhizosphere soil microbial index (RSMI) and found dehydrogenase, BSR/MBC, MBC/SOC, EC, phenol oxidase, and active microbial biomass carbon (AMBC) to be the most critical properties (Sinha et al. 2009). Comparing the performance of the three SQI developed as (i) simple additive SQI (SQI-1), (ii) weighted additive SQI (SQI-2), and (iii) statistically modelled SQI (SQI-3) based on PCA reported that SQI-3 was most strongly correlated (r = 0.74–0.78) with crop yield (Mukherjee and Lal 2014). A PCA-based Technosols quality indicator showed DHA, available nitrogen (N), and silt content were the key soil parameters affected by mine reclamation using tree species mixture and sufficient to indicate mine soil quality (Ahirwal et al. 2018). Based on indicator parameters, authors reported higher Technosols quality index (TQI) values for reclaimed mine soils compared to forest soils. Similarly, the study of microbial indicators shows that denitrification, respiration, and its ratio act as bioindicators of soil functioning after the disturbance caused by gold mining in tropical rainforests of French Guiana (Schimann et al. 2012).

5.5 Overview

Mining activities alter the landscape structure and function of the ecosystem which has severe environmental consequences. Restoration of mined land can be carried out through plantation, soil amendments, and different management practices. Monitoring and assessment of the potential impacts of mining and restoration measures that perturb the soil quality are of high priority. Restoration strategies projected to evaluate the disturbance and restoration use a specific set of indicator parameters. The magnitude of recovery or impact of different management practices on ecosystem functions and processes must be determined using indicator parameters. Indicators of soil quality should be efficient to evaluate the impact of management practices and be widely accepted in different ecosystems. As there is no unanimously used indicator to determine the soil quality, there needs to be a standardization of the indicators that can be frequently used to assess the soil quality. Biological indicator parameters compared to physical and chemical parameters help to assess the recovery in the early stages of restoration.

References

Agus, C., Primananda, E., Faridah, E. et al. (2018). Role of arbuscular mycorrhizal fungi and Pongamia pinnata for revegetation of tropical open-pit coal mining soils. *International Journal of Environmental Science and Technology* 16 (7): 3365–3374. https://doi.org/10.1007/S13762-018-1983-5.

Ahirwal, J. and Maiti, S.K. (2018). Assessment of soil carbon pool, carbon sequestration and soil CO2 flux in unreclaimed and reclaimed coal mine spoils. *Environmental Earth Sciences* 77: 9. https://doi.org/10.1007/s12665-017-7185-5.

Ahirwal, J., Maiti, S.K., and Singh, A.K. (2016). Ecological restoration of coal mine-degraded lands in dry tropical climate: what has been done and what needs to be done? *Environmental Quality Management* 26: 25–36. https://doi.org/10.1002/TQEM.21481.

Ahirwal, J., Maiti, S.K., and Satyanarayana, R.M. (2017). Development of carbon, nitrogen and phosphate stocks of reclaimed coal mine soil within 8 years after forestation with Prosopis juliflora (Sw.) Dc. *Catena* 156: 42–50. https://doi.org/10.1016/j.catena.2017.03.019.

Ahirwal, J., Kumar, A., Pietrzykowski, M., and Maiti, S.K. (2018). Reclamation of coal mine spoil and its effect on Technosol quality and carbon sequestration: a case study from India. *Environmental Science and Pollution Research* 25: 27992–28003. https://doi.org/10.1007/s11356-018-2789-1.

Ahirwal, J., Kumar, A., and Maiti, S.K. (2020). Effect of fast-growing trees on soil properties and carbon storage in an afforested coal mine land (India). *Minerals* 10: 1–15. https://doi.org/10.3390/min10100840.

Ahirwal, J., Kumari, S., Singh, A.K. et al. (2021). Changes in soil properties and carbon fluxes following afforestation and agriculture in tropical forest. *Ecological Indicators* 123: https://doi.org/10.1016/j.ecolind.2021.107354.

Alday, J.G., Santana, V.M., Marrs, R.H., and Martínez-Ruiz, C. (2014). Shrub-induced understory vegetation changes in reclaimed mine sites. *Ecological Engineering* 73: 691–698. https://doi.org/10.1016/J.ECOLENG.2014.09.079.

Allison, S.D. and Vitousek, P.M. (2005). Responses of extracellular enzymes to simple and complex nutrient inputs. *Soil Biology and Biochemistry* 37: 937–944. https://doi.org/10.1016/J.SOILBIO.2004.09.014.

Andrews, S.S. and Carroll, C.R. (2001). Designing a soil quality assessment tool for sustainable agroecosystem management. *Ecological Applications* 11: 1573–1585.

Asensio, V., Vega, F.A., and Covelo, E.F. (2014). Effect of soil reclamation process on soil C fractions. *Chemosphere* 95: 511–518. https://doi.org/10.1016/J.CHEMOSPHERE.2013.09.108.

Baldrian, P., Trögl, J., Frouz, J. et al. (2008). Enzyme activities and microbial biomass in topsoil layer during spontaneous succession in spoil heaps after brown coal mining. *Soil Biology and Biochemistry* 40: 2107–2115. https://doi.org/10.1016/j.soilbio.2008.02.019.

Banning, N.C., Grant, C.D., Jones, D.L., and Murphy, D.V. (2008). Recovery of soil organic matter, organic matter turnover and nitrogen cycling in a post-mining forest rehabilitation chronosequence. *Soil Biology and Biochemistry* 40: 2021–2031. https://doi.org/10.1016/j.soilbio.2008.04.010.

Barré, P., Durand, H., Chenu, C. et al. (2017). Geological control of soil organic carbon and nitrogen stocks at the landscape scale. *Geoderma* 285: 50–56. https://doi.org/10.1016/J.GEODERMA.2016.09.029.

Bujalský, L., Kaneda, S., Dvorščík, P., and Frouz, J. (2014). In situ soil respiration at reclaimed and unreclaimed post-mining sites: responses to temperature and reclamation treatment. *Ecological Engineering* 68: 53–59. https://doi.org/10.1016/j.ecoleng.2014.03.048.

Buta, M., Blaga, G., Paulette, L. et al. (2019). Soil reclamation of abandoned mine lands by revegetation in Northwestern part of Transylvania: a 40-year retrospective study. *Sustainability* 11: 3393. https://doi.org/10.3390/SU11123393.

Chodak, M. and Niklińska, M. (2010a). The effect of different tree species on the chemical and microbial properties of reclaimed mine soils. *Biology and Fertility of Soils* 46: 555–566. https://doi.org/10.1007/s00374-010-0462-z.

Chodak, M. and Niklińska, M. (2010b). Effect of texture and tree species on microbial properties of mine soils. *Applied Soil Ecology* 46: 268–275. https://doi.org/10.1016/j.apsoil.2010.08.002.

Ciarkowska, K., Sołek-Podwika, K., and Wieczorek, J. (2014). Enzyme activity as an indicator of soil-rehabilitation processes at a zinc and lead ore mining and processing area. *Journal of Environmental Management* 132: 250–256. https://doi.org/10.1016/J.JENVMAN.2013.10.022.

Dhar, A., Comeau, P.G., Naeth, M.A. et al. (2020). Plant community development following reclamation of oil sands mines using four cover soil types in northern Alberta. *Restoration Ecology* 28: 82–92. https://doi.org/10.1111/REC.13039.

Dhawi, F., Datta, R., and Ramakrishna, W. (2015). Mycorrhiza and PGPB modulate maize biomass, nutrient uptake and metabolic pathways in maize grown in mining-impacted soil. *Plant Physiology and Biochemistry* 97: 390–399. https://doi.org/10.1016/J.PLAPHY.2015.10.028.

Dick, R.P. (1997). Soil enzyme activities as integrative indicators of soil health. In: *Biological Indicators of Soil Health* (ed. C. Pankhurst, B.M. Doube and V.V.S.R. Gupta), 121–156. New York: CAB International.

Feng, Y., Wang, J., Bai, Z., and Reading, L. (2019). Effects of surface coal mining and land reclamation on soil properties: a review. *Earth-Science Reviews* 191: 12–25. https://doi.org/10.1016/j.earscirev.2019.02.015.

Finkenbein, P., Kretschmer, K., Kuka, K. et al. (2013). Soil enzyme activities as bioindicators for substrate quality in revegetation of a subtropical coal mining dump. *Soil Biology and Biochemistry* 56: 87–89. https://doi.org/10.1016/j.soilbio.2012.02.012.

Frost, S.M., Stahl, P.D., and Williams, S.E. (2001). Long-term reestablishment of arbuscular mycorrhizal fungi in a drastically disturbed semiarid surface mine soil. *Arid Land Research and Management* 15: 3–12. https://doi.org/10.1080/15324980119429.

Frouz, J. and Bujalský, L. (2018). Flow of CO_2 from soil may not correspond with CO_2 concentration in soil. *Scientific Reports* 8: 1–10. https://doi.org/10.1038/s41598-018-28225-z.

Ghosh, D. and Maiti, S.K. (2021). Effect of invasive weed biochar amendment on soil enzymatic activity and respiration of coal mine spoil: a laboratory experiment study. *Biochar* 3 (4): 519–533. https://doi.org/10.1007/S42773-021-00109-Y.

Goupil, K. and Nkongolo, K. (2014). Assessing soil respiration as an indicator of soil microbial activity in reclaimed metal contaminated lands. *American Journal of Environmental Sciences* 10: 403–411. https://doi.org/10.3844/AJESSP.2014.403.411.

Graham, M.H. and Haynes, R.J. (2004). Organic matter status and the size, activity and metabolic diversity of the soil microflora as indicators of the success of rehabilitation of mined sand dunes. *Biology and Fertility of Soils* 39 (6): 429–437. https://doi.org/10.1007/S00374-003-0713-3.

Haney, R.L., Hossner, L.R., and Haney, E.B. (2008). Soil microbial respiration as a tool to assess post mine reclamation. *International Journal of Mining, Reclamation and Environment* 22: 48–59. https://doi.org/10.1080/17480930701414584.

Hinojosa, M.B., Carreira, J.a., García-Ruíz, R., and Dick, R.P. (2004). Soil moisture pre-treatment effects on enzyme activities as indicators of heavy metal-contaminated and reclaimed soils. *Soil Biology and Biochemistry* 36: 1559–1568. https://doi.org/10.1016/j.soilbio.2004.07.003.

Ingram, L.J., Schuman, G.E., Stahl, P.D., and Spackman, L.K. (2005). Microbial respiration and organic carbon indicate nutrient cycling recovery in reclaimed soils. *Soil Science Society of America Journal* 69: 1737–1745. https://doi.org/10.2136/sssaj2004.0371.

Izquierdo, I., Caravaca, F., Alguacil, M.M. et al. (2005). Use of microbiological indicators for evaluating success in soil restoration after revegetation of a mining area under subtropical conditions. *Applied Soil Ecology* 30: 3–10. https://doi.org/10.1016/j.apsoil.2005.02.004.

Juwarkar, A.A., Mehrotraa, K.L., Nair, R. et al. (2010). Carbon sequestration in reclaimed manganese mine land at Gumgaon, India. *Environmental Monitoring and Assessment* 160: 457–464. https://doi.org/10.1007/s10661-008-0710-y.

Karlen, D.L., Mausbach, M.J., Doran, J.W. et al. (1997). Soil quality: a concept, definition, and framework for evaluation (a Guest editorial). *Soil Science Society of America Journal* 61: 4–10. https://doi.org/10.2136/SSSAJ1997.03615995006100010001X.

Kocira, A., Staniak, M., Tomaszewska, M. et al. (2020). Legume cover crops as one of the elements of strategic weed management and soil quality improvement. A review. *Agriculture* 10: 394. https://doi.org/10.3390/AGRICULTURE10090394.

Kumar, S., Maiti, S.K., and Chaudhuri, S. (2015). Soil development in 2-21 years old coalmine reclaimed spoil with trees: a case study from Sonepur-Bazari opencast project, Raniganj Coalfield, India. *Ecological Engineering* 84: 311–324. https://doi.org/10.1016/j.ecoleng.2015.09.043.

Kumari, S. and Maiti, S.K. (2019). Reclamation of coalmine spoils with topsoil, grass, and legume: a case study from India. *Environmental Earth Sciences* 78: 429. https://doi.org/10.1007/s12665-019-8446-2.

Li, C., Shi, L.L., Ostermann, A. et al. (2015a). Indigenous trees restore soil microbial biomass at faster rates than exotic species. *Plant and Soil* 396 (1): 151–161. https://doi.org/10.1007/S11104-015-2570-X.

Li, J., Zhou, X., Yan, J. et al. (2015b). Effects of regenerating vegetation on soil enzyme activity and microbial structure in reclaimed soils on a surface coal mine site. *Applied Soil Ecology* 87: 56–62. https://doi.org/10.1016/J.APSOIL.2014.11.010.

Li, J., Li, H., Zhang, Q. et al. (2019). Effects of fertilization and straw return methods on the soil carbon pool and CO_2 emission in a reclaimed mine spoil in Shanxi Province, China. *Soil and Tillage Research* 195: 104361. https://doi.org/10.1016/J.STILL.2019.104361.

Liao, M. and Xie, X.M. (2007). Effect of heavy metals on substrate utilization pattern, biomass, and activity of microbial communities in a reclaimed mining wasteland of red soil area. *Ecotoxicology and Environmental Safety* 66: 217–223. https://doi.org/10.1016/J.ECOENV.2005.12.013.

Maiti, S.K. (2007). Bioreclamation of coalmine overburden dumps – with special empasis on micronutrients and heavy metals accumulation in tree species. *Environmental Monitoring and Assessment* 125: 111–122. https://doi.org/10.1007/s10661-006-9244-3.

Maiti, S.K. (2013). *Ecorestoration of Coal Mine Degraded Land*. Springer.

Maiti, S.K. and Ahirwal, J. (2018). Ecological restoration of coal mine degraded lands: topsoil management, pedogenesis, carbon sequestration, and mine pit limnology. In: *Phytomanagement of Polluted Sites: Market Opportunities in Sustainable Phytoremediation* (ed. V.C. Pandey and K. Bauddh), 83–111. Elsevier B.V https://doi.org/10.1016/B978-0-12-813912-7.00003-X.

Maiti, S.K., Kumar, A., and Ahirwal, J. (2016). Bioaccumulation of metals in timber and edible fruit trees growing on reclaimed coal mine overburden dumps. *International Journal of Mining, Reclamation and Environment* 30: 231–244. https://doi.org/10.1080/17480930.2015.1038864.

Malik, A. and Scullion, J. (1998). Soil development on restored opencast coal sites with particular reference to organic matter and aggregate stability. *Soil Use and Management* 14: 234–239. https://doi.org/10.1111/J.1475-2743.1998.TB00158.X.

Masto, R.E., Ram, L.C., Shandilya, P.R. et al. (2011). Selection of bioindicators in coal-contaminated soils of Dhanbad, India. *Environmental Earth Sciences* 64: 1107–1115. https://doi.org/10.1007/s12665-011-0927-x.

Mastrogianni, A., Papatheodorou, E.M., Monokrousos, N. et al. (2014). Reclamation of lignite mine areas with *Triticum aestivum*: the dynamics of soil functions and microbial communities. *Applied Soil Ecology* 80: 51–59.

Mathiba, M., Awuah-Offei, K., and Baldassare, F.J. (2014). Influence of elevation, soil temperature and soil moisture content on reclaimed mine land soil CO_2 fluxes. *Environmental Earth Sciences* 73 (10): 6131–6143. https://doi.org/10.1007/S12665-014-3839-8.

Memoli, V., De Marco, A., Esposito, F. et al. (2019). Seasonality, altitude and human activities control soil quality in a national park surrounded by an urban area. *Geoderma* 337: 1–10. https://doi.org/10.1016/J.GEODERMA.2018.09.009.

Monokrousos, N., Boutsis, G., and Diamantopoulos, J.D. (2014). Development of soil chemical and biological properties in the initial stages of post-mining deposition sites. *Environmental Monitoring and Assessment* 186 (12): 9065–9074. https://doi.org/10.1007/S10661-014-4066-1.

Mukherjee, A. and Lal, R. (2014). Comparison of soil quality index using three methods. *PloS One* 9: e105981. https://doi.org/10.1371/journal.pone.0105981.

Mukhopadhyay, S. and Masto, R.E. (2016). Carbon storage in coal mine spoil by *Dalbergia sissoo* Roxb. *Geoderma* 284: 204–213. https://doi.org/10.1016/j.geoderma.2016.09.004.

Mukhopadhyay, S., Maiti, S.K., and Masto, R.E. (2013). Use of Reclaimed Mine Soil Index (RMSI) for screening of tree species for reclamation of coal mine degraded land. *Ecological Engineering* 57: 133–142. https://doi.org/10.1016/j.ecoleng.2013.04.017.

Mukhopadhyay, S., Maiti, S.K., and Masto, R.E. (2014). Development of mine soil quality index (MSQI) for evaluation of reclamation success: a chronosequence study. *Ecological Engineering* 71: 10–20. https://doi.org/10.1016/j.ecoleng.2014.07.001.

Mukhopadhyay, S., Masto, R.E., Cerdà, A., and Ram, L.C. (2016). Rhizosphere soil indicators for carbon sequestration in a reclaimed coal mine spoil. *Catena* 141: 100–108. https://doi.org/10.1016/j.catena.2016.02.023.

Mummey, D.L., Stahl, P.D., and Buyer, J.S. (2002). Microbial biomarkers as an indicator of ecosystem recovery following surface mine reclamation. *Applied Soil Ecology* 21: 251–259.

Muñoz-Rojas, M., Erickson, T.E., Dixon, K.W., and Merritt, D.J. (2016). Soil quality indicators to assess functionality of restored soils in degraded semiarid ecosystems. *Restoration Ecology* 24: S43–S52. https://doi.org/10.1111/REC.12368.

Novak, J.M., Ippolito, J.A., Ducey, T.F. et al. (2018). Remediation of an acidic mine spoil: Miscanthus biochar and lime amendment affects metal availability, plant growth, and soil enzyme activity. *Chemosphere* 205: 709–718. https://doi.org/10.1016/J.CHEMOSPHERE.2018.04.107.

Palmer, C.J., Smith, W.D., and Conkling, B.L. (2002). Development of a protocol for monitoring status and trends in forest soil carbon at a national level. *Environmental Pollution* 116: S209–S219. https://doi.org/10.1016/S0269-7491(01)00253-6.

Pietrzykowski, M. and Daniels, W.L. (2014). Estimation of carbon sequestration by pine (Pinus sylvestris L.) ecosystems developed on reforested post-mining sites in Poland on differing mine soil substrates. *Ecological Engineering* 73: 209–218. https://doi.org/10.1016/j.ecoleng.2014.09.058.

Piotrowska-Seget, Z., Cycoń, M., and Kozdrój, J. (2005). Metal-tolerant bacteria occurring in heavily polluted soil and mine spoil. *Applied Soil Ecology* 28: 237–246. https://doi.org/10.1016/j.apsoil.2004.08.001.

Qian, K., Wang, L., and Yin, N. (2012). Effects of AMF on soil enzyme activity and carbon sequestration capacity in reclaimed mine soil. *International Journal of Mining Science and Technology* 22: 553–557. https://doi.org/10.1016/J.IJMST.2012.01.019.

de Quadros, P.D., Zhalnina, K., Davis-Richardson, A.G. et al. (2016). Coal mining practices reduce the microbial biomass, richness and diversity of soil. *Applied Soil Ecology* 98: 195–203. https://doi.org/10.1016/J.APSOIL.2015.10.016.

Rumpel, C., Kögel-Knabner, I., and Hüttl, R.F. (1999). Organic matter composition and degree of humification in lignite-rich mine soils under a chronosequence of pine. *Plant and Soil* 213 (1): 161–168. https://doi.org/10.1023/A:1004454826537.

Schimann, H., Petit-jean, C., Guitet, S. et al. (2012). Microbial bioindicators of soil functioning after disturbance : The case of gold mining in tropical rainforests of French Guiana. *Ecological Indicators* 20: 34–41. https://doi.org/10.1016/j.ecolind.2012.01.021.

Sencindiver, J.C. and Ammons, J.T. (2000). Minesoil genesis and classification. In: *Reclamation of Drastically Disturbed Lands* (ed. R.I. Barnhisel, R.G. Darmody and W.L. Daniels), 595–613. John Wiley & Sons, Ltd https://doi.org/10.2134/AGRONMONOGR41.C23.

Shrestha, P., Gautam, R., and Ashwath, N. (2019). Effects of agronomic treatments on functional diversity of soil microbial community and microbial activity in a revegetated coal mine spoil. *Geoderma* 338: 40–47. https://doi.org/10.1016/J.GEODERMA.2018.11.038.

Shukla, M.K. and Lal, R. (2005). Soil organic carbon stock for reclaimed minesoils in northeastern Ohio. *Land Degradation and Development* 16: 377–386. https://doi.org/10.1002/ldr.669.

Singh, J.S., Raghubanshi, A.S., Singh, R.S., and Srivastava, S.C. (1989). Microbial biomass acts as a source of plant nutrients in dry tropical forest and savanna. *Nature* 338 (6215): 499–500. https://doi.org/10.1038/338499a0.

Singh, a.N., Raghubanshi, a.S., and Singh, J.S. (2004). Impact of native tree plantations on mine spoil in a dry tropical environment. *Forest Ecology and Management* 187: 49–60. https://doi.org/10.1016/S0378-1127(03)00309-8.

Sinha, S., Masto, R.E., Ram, L.C. et al. (2009). Soil Biology & Biochemistry Rhizosphere soil microbial index of tree species in a coal mining ecosystem. *Soil Biology and Biochemistry* 41: 1824–1832. https://doi.org/10.1016/j.soilbio.2008.11.022.

Song, Z., Bi, Y., Zhang, J. et al. (2020). Arbuscular mycorrhizal fungi promote the growth of plants in the mining associated clay. *Scientific Reports* 10 (1): 1–9. https://doi.org/10.1038/s41598-020-59447-9.

Soria, R., Ortega, R., Bastida, F., and Miralles, I. (2021). Role of organic amendment application on soil quality, functionality and greenhouse emission in a limestone quarry from semiarid ecosystems. *Applied Soil Ecology* 164: 103925. https://doi.org/10.1016/J.APSOIL.2021.103925.

Šourková, M., Frouz, J., and Šantrůčková, H. (2005). Accumulation of carbon, nitrogen and phosphorus during soil formation on alder spoil heaps after brown-coal mining, near Sokolov (Czech Republic). *Geoderma* 124: 203–214. https://doi.org/10.1016/j.geoderma.2004.05.001.

Srivastava, S.C. and Singh, J.S. (1989). Changes with time in soil biomass C, N and P of mine spoils in a dry tropical environment. *Canadian Journal of Soil Science* 69: 849–855.

Tang, J., Zhang, L., Zhang, J. et al. (2020). Physicochemical features, metal availability and enzyme activity in heavy metal-polluted soil remediated by biochar and compost. *Science of The Total Environment* 701: 134751. https://doi.org/10.1016/J.SCITOTENV.2019.134751.

Tordoff, G.M., Baker, A.J.M., and Willis, A.J. (2000). Current approaches to the revegetation and reclamation of metalliferous mine wastes. *Chemosphere* 41: 219–228. https://doi.org/10.1016/S0045-6535(99)00414-2.

Ussiri, D.A.N., Lal, R., and Jacinthe, P.A. (2006). Soil properties and carbon sequestration of afforested pastures in reclaimed minesoils of Ohio. *Soil Science Society of America Journal* 1797–1806. https://doi.org/10.2136/sssaj2005.0352.

Uzarowicz, Ł., Wolińska, A., Błońska, E. et al. (2020). Technogenic soils (Technosols) developed from mine spoils containing Fe sulphides: microbiological activity as an indicator of soil development following land reclamation. *Applied Soil Ecology* 156: 103699. https://doi.org/10.1016/ and enzyme activity in heavy metalJ.APSOIL.2020.103699.

Valente, F.D.A., Gomes, L.C., Castro, M.F. et al. (2021). Influence of different tree species on autotrophic and heterotrophic soil respiration in a mined area under reclamation. *Land Degradation & Development* 32: 4288–4299. https://doi.org/10.1002/LDR.4035.

Valle, S.R. and Carrasco, J. (2018). Soil quality indicator selection in Chilean volcanic soils formed under temperate and humid conditions. *CATENA* 162: 386–395. https://doi.org/10.1016/J.CATENA.2017.10.024.

Wang, F. (2017). Occurrence of arbuscular mycorrhizal fungi in mining-impacted sites and their contribution to ecological restoration: mechanisms and applications. *Critical Reviews in Environmental Science and Technology* 47: 1901–1957. https://doi.org/10.1080/10643389.2017.1400853.

Wang, Q., He, M., and Wang, Y. (2010). Influence of combined pollution of antimony and arsenic on culturable soil microbial populations and enzyme activities. *Ecotoxicology* 20 (1): 9–19. https://doi.org/10.1007/S10646-010-0551-7.

Xiao, L., Bi, Y., Du, S. et al. (2019). Effects of re-vegetation type and arbuscular mycorrhizal fungal inoculation on soil enzyme activities and microbial biomass in coal mining subsidence areas of Northern China. *CATENA* 177: 202–209. https://doi.org/10.1016/J.CATENA.2019.02.019.

Yang, S.X., Liao, B., Yang, Z.H. et al. (2016). Revegetation of extremely acid mine soils based on aided phytostabilization: a case study from southern China. *Science of The Total Environment* 562: 427–434. https://doi.org/10.1016/J.SCITOTENV.2016.03.208.

Yang, Y., Reilly, E.C., Jungers, J.M. et al. (2019). Climate benefits of increasing plant diversity in perennial bioenergy crops. *One Earth* 1: 434–445. https://doi.org/10.1016/J.ONEEAR.2019.11.011.

Zeng, P., Guo, Z., Xiao, X., and Peng, C. (2019). Effects of tree-herb co-planting on the bacterial community composition and the relationship between specific microorganisms and enzymatic activities in metal(loid)-contaminated soil. *Chemosphere* 220: 237–248. https://doi.org/10.1016/J.CHEMOSPHERE.2018.12.073.

Zhan, J. and Sun, Q. (2014). Development of microbial properties and enzyme activities in copper mine wasteland during natural restoration. *CATENA* 116: 86–94. https://doi.org/10.1016/J.CATENA.2013.12.012.

Zhang, F.P., Li, C.F., Tong, L.G. et al. (2010). Response of microbial characteristics to heavy metal pollution of mining soils in central Tibet, China. *Applied Soil Ecology* 45: 144–151. https://doi.org/10.1016/J.APSOIL.2010.03.006.

Zhao, R., Guo, W., Bi, N. et al. (2015). Arbuscular mycorrhizal fungi affect the growth, nutrient uptake and water status of maize (Zea mays L.) grown in two types of coal mine spoils under drought stress. *Applied Soil Ecology* 88: 41–49. https://doi.org/10.1016/J.APSOIL.2014.11.016.

Zhijun, H., Selvalakshmi, S., Vasu, D. et al. (2018). Identification of indicators for evaluating and monitoring the effects of Chinese fir monoculture plantations on soil quality. *Ecological Indicators* 93: 547–554. https://doi.org/10.1016/J.ECOLIND.2018.05.034.

6

Ecosystem Services on Restored Mine Land

Abstract

Destruction of natural ecosystems due to mining cannot be averted as soil and vegetation are lost. However, restoration of mine lands can contribute toward ecosystem remodeling, through planning, systematically identifying goals and objectives. A restoration planning model is applicable on degraded topsoils where practical consideration is prior. In this regard, model examples and case studies of restored sites can be explored to solve critical mine site problems. With the development of ecological restoration techniques through mine land reclamation and revegetation, loss of biodiversity can be restored to some extent. Although the ecological theory is deficient in general ordinance with application at the ecosystem level, knowledge about the ecosystem has applicability to targeted restoration goals through essential supervision and management irrespective of the apprehensions in restoration planning. In this regard, policy restructuring through proper planning techniques of ecosystem-based restoration can open up new avenues of technology-based restoration models. The idea of circular economy can be roped in with mine land restoration for a sustainable future, through proper management and monitoring at the post-mining stages. This chapter discusses the ecosystem services from rehabilitated mine land and emphasizes on a circular economy-based mine land restoration.

Keywords *Provisional services; regulatory services; supporting services; cultural services; coal mine spoil; restoration*

Eco-Restoration of Mine Land, First Edition. Vimal Chandra Pandey, Jitendra Ahirwal, Roopali Roychowdhury, and Ritu Chaturvedi.

CONTENTS

6.1 Introduction

Mines left abandoned or "orphan" mines that are not in use are unsafe. If the mine is left unattended and not revamped, then the leachates of the mine will continue to degrade the surrounding landscapes. However, this process of rehabilitation of a mine is not simple. As each mine is different so shall be its rehabilitation needs. Generally, mines containing heavy metals and ores tend to leave behind hazardous chemicals which turn into acids in contact with water or rain causing serious effects if exposed. Sulfur mines are particular in this aspect. If these individual mines are not closed, further oxidization of the leftover chemicals will continue to harm the air quality, adjacent areas, and would affect local communities.

For other mines, the strategies are different. Based on the nature of erosion, and land itself, a need for new soil is to be laid out, along with plantation of new vegetation, to restore the pre-mining land conditions. Rehabilitation requires structural changes which would help the land to refurbish without harming the local surroundings. However, prior local and commissioned planning by policy makers and stakeholders should be considered before restoration.[1]

Recommended post-mining land use is what we know as mine reclamation or rehabilitation. A large area of these lands can be rehabilitated with proper techniques if not in total. However, total reclamation is possible using advanced approaches, the success rate of which is also quite promising in various countries. Prior mining, companies formulate a restoration plan in consortium with other stakeholders (Sánchez et al. 2014; ICMM 2019) taking into account different sets of inconsistencies and criteria, including subsurface water mark, end pit, and waste pile formations, afforestation plans, stability of land its use and policies, and the original land environment. Thus, due to mining's effect on natural or cultivated environs such as forests, an array of essential gains termed as ecosystem services are either lost or adversely affected (Rosa and Sánchez 2016). In similar positions, the main motive in favor of restoration should be focused to obtain preconditions of mine lands wherever possible. Societal benefits are emphasized as a result of rehabilitation efforts when ecosystem services are taken into consideration during planning and evaluation (Rosa et al. 2018). Local community needs and concerns along with environment protection if prioritized before framing modules for land-use planning can be promising for ecosystem services (Baral et al. 2016; Geneletti 2016; Jacob et al. 2016; Lazoz-Chavero et al. 2016; Cortinovis and Geneletti 2018; Rozas-Vásquez et al. 2018). Re-constructing stipulated ecosystem services via edible crop production on agricultural land or forest-based goods has been documented by Larondelle and Haase (2012), and Zipper et al. (2011) as positive examples in mine land reclamation. Mining activities provide compulsory modifications toward the ecosystem which affects its potential benefits. These are termed ecosystem services. Since 2006, assessing impacts on ecosystem services has been essential. Post-mine closure, rehabilitation of the degraded sites and their utilization to pre-mining times for biodiversity and other utilities are required for operation. This restoration of ecosystem-based services or utilities is a tedious task and depends largely on the quality and output of these functions (Maestre et al. 2012).

Ecosystem services (ES) remain a significant and obscure path to frame the environmental and communal consequences of quarrying (Rosa and Sánchez 2015) and rehabilitate it to count upon its potential as reforestation is reduced, the more an ecosystem is devastated (Bullock et al. 2011). However, the efficacy of rehabilitation remains underdiagnosed, which was documented by Rey-Benayas et al. (2009) through meta-analysis which revealed environmental restoration upgraded diversity as 44%, while the amenities of ecosystem services was 25%.

Rehabilitation monitoring mostly includes common ecological indicators such as species variance, abundance, and vegetation cover (Clements et al. 2010); however, a sustainable after mining condition is not always accounted by a typical fauna or flora species (Harris and van Diggelen 2006) as these indicators are inadequate for restoration mapping as social benefits by the restored sites remain untraced. It has been observed that many mining sites affect the cultural ecosystem services (Petrova and Marinova 2015) in various places; however, they are always little considered in terms of rehabilitation. Mostly, the observed data on vegetation cover or species diversity are taken into consideration rather than socio-cognitive measures (Aronson et al. 2010). The assumptions about mine land reclamation to restore all ecosystem services (Koch and Hobbs 2007; Rosa et al. 2018) remain unnoticed and unobserved and lack demonstration, but such end results are seldom tracked and cannot be readily manifested. Different cosystem services require a separate approach and exchange within each other (Geneletti 2013; Bullock et al. 2018). Earlier, settlements between cultural and provisioning services have been documented (Howe et al. 2014; Geneletti et al. 2018). Of late, the cultural ecosystem is being considered for evaluation (Ungaro et al. 2016; Jaligot et al. 2018) as it serves spiritual upliftment, rational development, and a variety of benefits for the locals along with an appraisal of the scenic beauty (Chan and Satterfield 2016). Assessment for cultural services can be done via fiscal (Polizzi et al. 2015; Mayer and Woltering 2018) and non-fiscal (Darvill and Lindo 2015; Bryce et al. 2016; Bullock et al. 2018) processes.

Being immaterial and intangible, cultural services are painstaking to be assessed compared to physical services. Evaluation based on participation and engagement of stakeholders, their point of measurement of value adds, development of opinion polls, review, and talks to understand benefits at the local level has to be done for the assessment process (Hermes et al. 2018; Raum 2018). Additionally, spatial study (Darvill and Lindo 2015) and internet community research (Orsi and Geneletti 2013; Richards and Tunçer 2018) have largely been utilized in collaboration with the above.

6.2 Rehabilitated Mine Land

Mining not only disturbs the mine land but also its surroundings, which include both underground and strip mining. While the former involves digging and tunneling the ground for deep excavation activities unfolding dirt and soil to get coal and other mineral ores, the latter removes land to obtain minerals nearer to the surface. Due to these activities, the land suffers deterioration. Soil disruption causes erosion due to vegetation loss and is further abused by rain, storms, hurricanes, or other calamities of nature leading to complete devastation of mine lands.[2] On the other hand, left out or orphan mines face hazards of leaching of metals and other compounds which under environmental influence damage entire land and water and disrupt local vegetation or agriculture. Due to these unwanted hazards, the remediation of mine lands becomes the utmost priority to keep away with the negative impacts on soil and water. Post-mining sites require restoration of the ecosystem as a common threshold. It is reported that mine digging impacts near about 1% of the land surface (Walker 1999); however, including post-industrial sites including ash and other wastes, the figure will surge. The strong impacts of mining on the ecosystem cannot be denied along with critical environmental degradation. The detrimental effects are highlighted by rising human health issues due to pollution of water, soil, and air (Palmer et al. 2010). The main perspective of the restoration lies in reducing the harmful effects of quarrying on the existing biosphere. As damaged mine sites have lost their resources and greenery, regeneration remains an ultimate but time-consuming affair. Presently, a large focus is being laid on restoring these devastated sites through different approaches. Restoration scientists are also looking into options of alternative ecosystem build-up which is capable enough to thrive upon degraded soil (Bradshaw 2000). Rehabilitation of devastated areas, through habitat restoration along with balancing, functioning, and service of the ecosystem structure, (SER 2004), is essential to counteract the loss of biological resources and to achieve ecological continuity of mining operations (Lawrence 2003; Suding 2011; Gastauer et al. 2013; Perring et al. 2015). As afforestation induces photosynthesis in these areas (Skousen 2010), along with litter deposition and its decomposition enhances the chances of carbon sequestration, nutrient cycling, and ecosystem functioning toward consistency (Sheoran et al. 2010; Cristescu et al. 2012; Ranjan et al. 2015). However, this process of reclamation remains critical which involves the choice, breeding, and reinstating of favorable species (Nussbaumer et al. 2016). Details about its ecology before using preferred plants and their growth management are essential (Native Plant Network 2016; Giannini et al. 2017) as hostile conditions and adverse interactions between choiced and untapped organisms may defer rehabilitation (Skousen 2010; Suding et al. 2015). Failure in mine site rehabilitation has taken place in some cases where lack of knowledge of invasive species has resulted in adverse effects on the native plant and animal life as well as impacted economy and human health (Elton 1958). Thus, the focus should be laid on monitoring-based restoration practices with definite restoration goals when trial and error procedures are adopted to seek out knowledge, especially in megadiverse tropical ecosystems (Barr et al. 2017).

6.2.1 Need for Mine Site Reclamation

- Huge excavation of land sites for ore and mineral recovery causes land deformities.
- Overlapping forests and mineral land cause damage to the adjacent agrizones that need a cautious approach.

- Land mass overburdening due to piling of wastes in and around the lease.
- As land remains the key factor in mining, sustainable development of the degraded site and surrounding dumps is essential.
- Mining involves segregation hence the impact of waste generation on the surrounding environment is alarming.
- Though pre-mining conditions are hard to restore, upgrading land by the development of recreation or water sport areas, plantation farms or forest cover is essential to reduce land wastage.

6.2.2 Reclamation of Mine Lands: From Spoils to Soils

Ecological restoration and ecological rehabilitation in mining are analogous as both relate to the mitigation of toxins and improvisations in terms of land stability, biodiversity, hydrological cycle, vegetation, soil fertility, and overall land development. This is a common feature in India and often in other parts of the world. However, rehabilitation usually requires an engineered approach. This includes backfilling mine voids, surface leveling and redefining crude rock, repairing drainage, preserving, nullifying toxic and other wastes and their removals, and change of surface soil to unite the water–soil cycles. Reclaimed mine land soils are often termed as "technosols" due to scientific efforts required (IUSS 2006). Although the efforts are high, the efficiency in context to soil fertility remains a challenge as mere translocation of soil would not render effective function of the water–soil cycle. For successful remodeling of soil, topsoil enhancement by addition of biochar from *Lantana camara* showed encouraging results in reconditioning soil quality and organic carbon content (Ghosh et al. 2020).

As ecological restoration starts after reintroduction of species in managed soils, careful selection of foliage with the ability to grow on disturbed and recuperating soils speeds up the successional process (Ahirwal et al. 2016). It is understood that ecological restoration begins as the following steps to manage soil structure and introduction of specific plant species for regeneration happen. However, the rehabilitation process speeds up as the succession of plant varieties and species diversification is visible in disturbed and devastated lands. Vegetation cover is generally speeded up by a careful selection of plantations suitable for quick growth on otherwise nutrition-less soils. In general, the afforestation process of mine lands produces semi-natural ecosystems that are anyways lacking biodiversity and relevant ecological functions and have bypassed many lineage stages (Ahirwal et al. 2016). This indicates the obstacles of undertaking ecological issues on mining sites.

India and other emerging nations have gained enough inspiration in addressing the issues of ecological rehabilitation of mined sites from different countries (Muñoz-Rojas et al. 2018; Ohsowski et al. 2018; Mukhopadhyay and Masto 2019). Careful selection of plant species on devastated coal mine land enhanced mine soil quality, and carbon content after a span of 16 years of revegetation was documented by Ahirwal and Maiti (2018). Further, the influence of microbiota in mine soils has become more compliant and practical (Li et al. 2019). Ecological rehabilitation of mined lands is speeded by early afforestation and tested the use of *Fabaceae* and *Poaceae* species to upgrade microbial growth in soil and geochemical cycling (Kumari and Maiti 2019). However, these are concerning mine sites having a different background, area conditions, topical and predicted weather, and climate conditions. Careful complemented rehabilitation plans and procedures are needed in each case, as studying the site is essential before implementing any ideology.

To align with the UN-SDGs, India can adopt better ecological rehabilitation practices in damaged mine sites. Though not always idealistic, community-based rehabilitation such as afforestation serves the purpose to quite an extent by remodeling land and generating employment at the local level thereby reducing poverty. In emerging nations during these difficult times of COVID-19, this paves a way toward economic stability and support for the underprivileged along with the rehabilitation of the disturbed and unproductive lands. Some of the alternatives like utilizing fields for energy production may not be quite feasible for India as more focus should be laid on local ecosystem enhancement, however, other possible approaches such as

- Fruit orchards development on mined land, along with scanning of toxicity (Maiti et al. 2016a).
- Bioenergy crops such as *Miscanthus* plantation on revamped mine soils (Ussiri et al. 2019).
- Solar energy plants are set up on refurbished mine lands for green energy and cutting down pollution levels in the environment (Mert 2019).

Although these efforts seem more of a hurricane task, nations actively involved in rehabilitation can focus on using better ecological techniques in their plan to achieve improvised ecological states and functions through people by involving them from local to national levels.

6.2.3 Reclamation Planning

The preliminary strategy of mine rehabilitation pinpoints such categories and consortiums who are already involved in similar work. Category of people involved in farming organizations and curators of customary land practices have important indigenous knowledge that can help minimize the unfavorable effects of mining and promote successful rehabilitation. The early phase in revegetation planning primarily deals in identifying knowledge gaps, and particular trials to derive critical information. In this regard, discussion with principle collaborator groups may give focused research and trial programs and upgrade the probable passing of knowledge into community projects. Moreover, each country, state, and territory has its lawful obligations related to grasping and administration of waste (spoil) materials at mine sites. Relevant authorities must be roped in by companies to share their requirements and any existing protocols that may need to be considered.

For durable rehabilitation of landforms, site-specific material characterization promotes selective positioning during landscape modelling to minimize risks of erosion or malfunctioning revegetation. Simultaneously, the rehabilitation, its approach, or initiatives will be prompt and economical. The effective design of the sampling protocol is directly related to the value of the scientific results used in the characterization of mine land materials; in this regard, researchers De Gruijter (2002), Yates and Warrick (2002), and Dollhopf (2000) have provided valuable inputs. Although research room testing is fruitful in identifying major challenges to constancy or plant growth, trials off site [glasshouse] (Asher et al. 2002) and site specific may also be validated to ascertain elusive aspects about the growth and accomplishment of different plant species in post-mining soils (Bell 2002). A geological examination is essential in characterizing mine piles, rock heaps, and refuses as the presence and identity of potentially acid-producing sulfides can be recognized which cause a deleterious effect on plant growth due to low pH values or via soluble metal concentrations. In this context, details of comprehensive suitable tests for assessing mineralogical parameters of mine sites for profitable plant growth are provided in soil mineralogy by Wilson (2003).

6.3 Ecosystem Services on Rehabilitated Coal Mine Land

The purpose of mine land rehabilitation is mainly biodiversity boost and enhanced services offered by the ecosystem (Whisenant 1999; Farber et al. 2006). In this regard, in the post-quarrying lands and industrial areas closed after mining are the pioneers for ecosystem services, concerning the physical parameters of erosion control, soil, water purification, and microbial diversification for consistent vegetation and ecosystem stability. However, in the later stages, focus can be laid on refurbishing the productive sites, carbon uptake, and aesthetics of the landscape. Given the above, ecologists accentuate the reformation of biodiversity as the ultimate step in restoration projects.

Biological diversity is a commendable norm in revegetation setup. However, they are not always specific as biodiversity indicators in mining site restorations target the choices which are value based and manageable (Pitz et al. 2016). To clarify innate alteration in plant habitat composition over a duration, the term "natural succession" is used. To naturally restore forest lands after devastation, new forest land in the former position is required. This credibility and rapidness entirely depend on the conditions left at the mine site post-mining (Groninger et al. 2017). Spontaneous succession (Prach and Pyšek 2001), or passive restoration (Holl and Aide 2011), depends primarily on natural succession for restoration of lands; however, technical restoration (Prach and Hobbs 2008; Holl and Aide 2011; Walker et al. 2014) is applied to disturbed landforms and involves remodeling of land through soil import and plantation activities, which may build up an alternate ecosystem which is useful and resolute (SER 2004), such as agroforestry (Metslaid et al. 2015). Abiotic conditions upgrade and restraining of irksome varieties are requisites of assisted restoration which includes planting or sowing of intent species.

If a land has lost its green cover including seeds and plant material effective of regrowth, and revegetation has not been initiated, in such places succession makes a slow move. Initially, grasses, herbs, vines, and shrubs colonize and rule the roost. These include goldenrods, ragweeds, and shrubs like raspberry and blackberry. However, the soil and site conditions determine their span or otherwise get replaced with other flora or forest trees. Other factors of soil conditions in mine sites also alter the speed of natural succession upon these areas, such as dug-out soils containing living seeds and roots from the former conditions can be applied to expedite succession. However, the mined areas that are near the forest will be populated by native forest species promptly in contrast to the farther ones (Groninger et al. 2017). This practice at mining and industrial sites aids in charging up the inherent ecosystem despite adverse conditions such as climate, nutrients, and pollution.

Significantly, mine soils are low in nutrients and high on metal concentration which need restoration through plant-based phytoremediation and phytostabilization methods for restoring metal toxic lands (Ali et al. 2013). As mentioned by Yang et al. (2015), woody species are better off to be used in metal-polluted lands than other herbaceous varieties as they live longer and have greater stress tolerance.

Reclaimed mine lands promote ecosystem augmentation through beneficial services toward mankind. Mine spoil lands and pits when revamped with soil and other organic wastes such as sewage, mud, farmyard manures, cattle wastes, and decomposed plants can prompt ecosystem build-up through afforestation techniques which would guarantee essential benefits in the form of provisional, social, cultural, and regulating services for humanity. A sustainable ecosystem and human wellness can be well targeted through this process and is also in demand nowadays. Ecosystem service is desired and is promoting estimation of social and ecological sustainability which can upgrade land-use planners and policy makers through socioeconomic knowledge.

6.3.1 Provisional Services

Restored mine land ecosystems provide a variety of services used by and for the betterment of mankind. Certain provisionary services which include timber, fodder, fuelwood, food products, organic compounds, and other valuable resources are offered in these areas. However, it is essential to evaluate the possible hazards of toxicity before utilizing food and fodder products. In addition to this, commercially valuable trees such as timber and fruits are also planted on these reclaimed mine lands. Different varieties of trees respond to metal stress through the accumulation of metals in their parts. Some timber coppice species such as *Leucaena leucocephala* (Lam.) de Wit, *Dalbergia sissoo* Roxb., *Acacia mangium* Willd., *Gmelina arborea* Roxb., *Bambusa arundanacea* L., and *Tectona grandis* L.f. have been reported by scientists as metal stabilizers and accumulators in coal digging sites (Maiti 2007; Maiti et al. 2016b). Flourishing trees such as *Melia azedarach* L. (Meliaceae) and *Acacia auriculiformis* A.Cunn.ex Benth. (Fabaceae) can also be used for revegetation (Das and Maiti 2016). As a prompt growing leguminous tree with successful growth rate *A. auriculiformis* is a tolerant of arid regions and has nitrogen-fixing ability (Peng et al. 2005), and *M. azedarach* is a rapidly growing tree with stress tolerance at infertile habitats (Maiti 2012). Afforestation activities on reclaimed mine lands endorse livelihood boost for people residing in the vicinity of mining sites. In this context, Shrivastava (2007) has reported the efficient role of tribals in tree plantation and their maintenance on fly ash dykes near Korba (India) which have developed as a new ecosystem through human intervention. Plants and plant-based products are greatly beneficial for people living in the vicinity of land mines.

However, bioaccumulation of toxins in vegetative parts also needs critical examination. Native trees and shrubs varieties such as oaks and hickory show rapid growth in mine soils (Hall et al. 2010) with favorable topsoil reclamation. These candidate varieties are edible sources and determine systematic attributes of flora on revamped mines. Different provisional services offered by natural and developed vegetation are thus useful in all aspects (Ahirwal et al. 2016).

6.3.2 Regulatory Services

Reclaimed mine land ecosystem also offers regulatory services which can be prioritized to administer pollution issues including metal leaching, soil erosion, and carbon regulation. Certain plant varieties are ecologically acclaimed and reported by researchers to fight pollution and regulate carbon sequestration through effective phytoremediation, (Mukhopadhyay and Maiti 2010) rhizoremediation, and stabilization techniques (Pandey et al. 2012). Forestation activities greatly influence soil biochemical properties as a result of litter fall and nutrient release by developed root biomass. Further, the presence of plant species *D. sissoo* and *L. lucocephala* which are nitrogen fixators contributes much to the soil quality after a prolonged period of 15 years. However, low microbiotic activities also hamper the N and P concentrations in the soil (Mukhopadhyay et al. 2013). Restoration or revegetation of damaged mine land prompts methane metabolism resulting in an increase of the methane pool of the spoilt soil and recovering it. Dense vegetation and plants would regulate temperature rise and absorb excess carbon dioxide build-up in the atmosphere.

Along with selected afforestation species, some ancillary tree species like *Azadirachta indica, Ficus religiosa, Prosopis juliflora, Ficus retusa, Zizyphus jujuba, Heloptelia integrifolia*, and *Trema orientalis* and, in reclamation sites were reported by Kumar et al. (2015b).

6.3.3 Supporting Services

The supporting services offered by mine lands after reclamation are none less. Rather, they are essential for the proper functioning of this eco-community and have a direct influence on the lives of people. Primarily, soil formation, nutrient cycling, photosynthesis, autotrophs, habitat, and adaptability are focused on supporting services. To support the local communities, some areas of revamped mine lands can be converted into parks and endeavor to promote the growth of biotic communities to initiate effective biogeochemical cycling. In this regard, flora development on these mine lands can regularize nutrient cycling, increase soil organic content, and develop microbial communities. Any type of vegetation on abandoned mine lands endorses soil remodeling. With a gradual increase in step-wise vegetation trials, an increase in organic debris with regularized litter fall, light soil, better permeability, enhanced anchorage of the root system, moisture retention, and a favorable microbial environment can be achieved. This would over a while stimulate the cropping of other plant varieties (Shukla and Misra 1993). In addition to conventional and naturally developed vegetation in these mine land sites, quite a few wildlife species that prefer young or mature forest conditions find ideal habitat conditions in these areas. Some species including amphibians, reptiles, birds, and mammals rely on immature forest covers for foraging, shelter, and breeding (Gilbart 2012). Native varieties such as vines, saplings, and shrubs which can flourish in less time are required to patch up spaces on mine lands to protect many declining wildlife species (Litvaitis 1993) by providing a survival habitat for them.

6.3.4 Cultural Services

Reclaimed mine lands provide impalpable and spiritual or intangible benefits for people by bestowing attractive and pleasing topography of the reclaimed damaged mine lands. These areas now impart natural greenery, cultural, spiritual, and religious improvement, traditional and scientific knowledge, support recreation, and ecotourism opportunities for the betterment of the natives as well as strengthen the economy by livelihood generation. Moreover, in the transition zones of rural and urban areas, recreational sports such as biking, car racing, soccer fields, outdoor sports, botanical gardens, public parks, auditoriums for outdoor activities as well as wildlife habitats can be initiated. In addition to this, reclaimed mine lands which have developed either through natural succession or through human intervention can be beautified either by planting aromatic grasses, biodiesel crops, flower gardens, or developing picnic spots for recreational purposes. Table 6.1 provides the details of ecosystem services offered by restored coal mine land.

Ecosystem services in reclaimed mine lands are highly influenced by plant–microbe interactions. To provide ecosystem-based services, efficient restoration of land is necessary. In this regard, the role of rhizomicrobes cannot be ignored as they have the potential of heavy metal absorption, plant growth promotion, and revitalizing restoration process. Metal-tolerant plant species such as *Jatropha* (Pandey et al. 2012) are effective producers of bio fuel; moreover, some other plant varieties secrete phytosiderophores which are effective in mineral sequestration, and thus the plant–microbe interaction can modify the biochemical and physical attributes of the soil substratum thereby benefiting the land ecosystem. Root secretions contain metabolites that help speed up the plant–microbe conglomerate. Thus, to accelerate the mine land ecosystem and associated services for mankind, utmost priority must be given to plant–microbe behavior and interactions.[3]

6.4 Restored Mine Land and United Nations Sustainable Development Goals

Land being a biophysical composite system dispenses crucial ecosystem services for mankind along with biodiversity maintenance, supporting nutrient cycling, and other ecological entities. It also remains an important factor for economic resources and human development along with sociocultural and religious importance.[4] Hence, maintaining the exuberance of the land resources is mandatory for the sake of the environment. Interestingly. climate change can be limited through land restoration while aiming for continuous development. The studies of the International Resource Panel (IRP) indicate positive outgrowths in support of the 2030 agenda for all the Sustainable Development Goals (SDGs). To achieve this

Table 6.1 Ecosystem services of restored coal mine lands.

Ecosystem service bracket	Products and services offered	Associated flora and fauna	References
Provisional services	Timber and fiber, Paper, Pulpwood, Construction, Furniture, Roof and flooring, Forage, Fruits/Food, Medicinal properties	*Eucalyptus spp.* *Shorea robusta* *Tectona grandis* *Grevillea robusta* *Bambusa sp.* *Acacia mangium* *Dalbergia sissoo* *Stylosanthes sp.* *Syzygium cumini* *Anacardium occidentale* *Psidium guajava* *Artocapus heterophyllus* *Mangifera indica* *Phyllanthus emblica*	Maiti et al. (2016b); Bandyopadhyay and Maiti (2019); Ahirwal and Maiti (2016); Maiti (2013); Maiti (2007); Kumari et al. (2022); Maiti et al. (2016b); Ranasinghe et al. (2019); Das and Maiti (2016)
Regulating services	Mitigation of pollution, Carbon sequestration, Eco-restoration and moisture retention capacity, Metal accumulation, Nitrogen fixation, Cover cropping, Soil quality improvement	*Acacia mangium* *Dalbergia sissoo* *Bambusa sp.* *Grevillea robusta* *Melia azedarach* *Acacia auriculiformis* *Stylosanthes hamata* *Sesbania sesban* *Acacia mangium* *Populus alba* *Crotalaria juncea*	Maiti et al. (2016b); Mukhopadhyay and Masto (2016); Rana and Maiti (2018); Das and Maiti (2016); Maiti et al. (2016a); Maiti and Maiti (2015)
Supporting Services	Surface runoff, prevention, Vegetative barrier, Agroforestry, Soil property, enhancement	*Acacia catechu* *Cymbopogon citratus* *Casuarina sp.* *Ficus religiosa* *Trema orientalis* *H. sabdariffa*	Maiti and Maiti (2015); Kumar et al. (2015b)
Cultural Services	Aesthetics, Energy and non-energy source, Recreational (public parks), Religious, Educational, (green capping and vegetation cover, organic farming)	*Lavandula sp.* *Helianthus annuus* *Miscanthus sp.* *Ficus religiosa* *Miscanthus sp.* *Cymbopogon citratus* *Casuarina sp.* *Lavandula sp.*	Chrysargyris et al. (2018); Kharytonov et al. (2019); Harris et al. (2015); Ussiri et al. (2019); Kumar et al. (2015a); Kharytonov et al. (2019); Maiti and Maiti (2015)

goal in 2030, immense effort is the need of the hour on all fronts. Presently, it is evaluated that almost a quarter of the global land is devastated.[5] A prudent and targeted restoration initiative will promote SDG targets linked to socio-economic development for instance, zero hunger, no poverty, gender equality, clean water and sanitation, good health and well-being, affordable energy production (bioenergy), decorous work, and economic uplift, reduced inequalities. Moreover, land restoration paves equal employment opportunities irrespective of gender and helps ending gender-biased jobs, and further supports female education[6] (Catacutan and Villamor 2016; Thompson-Hall 2016; Yurui et al. 2021).

Further, restoration procedures promote the financial improvement of women by altering devastated lands to agroforestry zones capable of food production (Robiglio and Reyes 2016). However, according to the United Nation Sustainable Development Goals (UN-SDGs),[7] the key SDGs have been categorized mainly under the following heads:

- no poverty
- zero hunger
- affordable and clean energy
- decent work and economic growth
- good health and well-being
- gender equality
- reduced inequalities
- clean water and sanitation
- responsible consumption and production
- climate action

Precisely, area [land] restoration has immense benefits and restoration has sheer support for other SDG targets also. The IRP of the UNEP has reported the UN-SDGs linked to land reclamation for which probable co-benefits are accessible. These are either linked head-on or incidentally, to the land quality which ensures life on land including care, revegetation, and continuous use of land ecosystems, forests, and mitigating biodiversity loss which relies on land resources.[8] Thus, as proposed for the "United Nations Decade on Ecosystem Restoration (UN-DER)," recovering land viability is needed for the appreciation of various intentions (UNEP 2015; Abhilash et al. 2016; Tripathi et al. 2017). Rapid changes in land resources due to pollution, urbanization, unsustainable land-use practices, deforestation, man-induced mining activities, overexploitation of water resources, and lack of policy measures have resulted in land resource shortage globally. Stocking defined land degradation as *temporary or permanent decline in the productive capacity of the land, and the diminution of the productive potential, including its major land, uses (e.g. rain-federal, irrigation, forests), its farming systems (e.g. smallholder subsistence), and its value as an economic resource* (Stockings 2001).

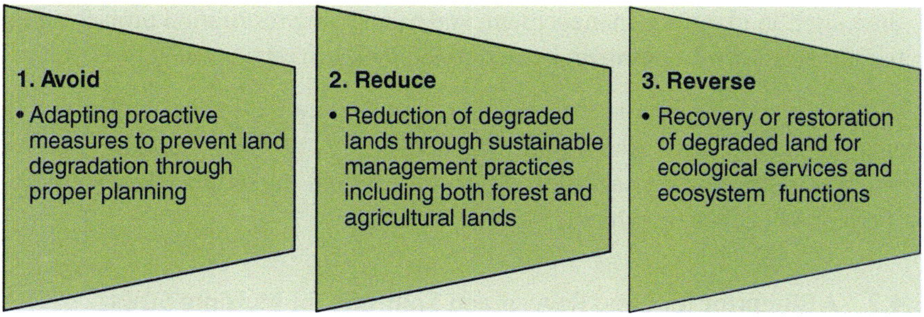

Figure 6.1 Three-step moto for sustainable restoration of lands.

Ecological complexities imparted by the land system should be the prime concern for the revegetation process of damaged lands (Syktus and McAlpine 2016). Thus, the critical moto of avoiding, reducing, and reversing land degradation during the UN-DER (2021–2030) and afar should be adopted. Figure 6.1 depicts the three-step moto for sustainable restoration of lands (Cowie et al. 2018; Orr et al. 2017).

On 25 September 2015, according to the ratification of the SDGs by the United Nations General Assembly, land restoration and rehabilitation, particularly "Life on Land" and "Land Degradation Neutrality," are the prime concerns of UNCCD's work has been worked upon.

6.4.1 Sustainable Development Goals Which Have Direct Benefits

- Poverty can be buckled down mainly by initiating the restoration process and promoting income and livelihood. Reduction of poor and vulnerable populations can be adopted as a step to restore the land which can ensure availability of water to the crops during dry periods and check runoffs causing floods in urban areas.
- Hunger correlates with subsistence farmers and ranchers, as they would benefit largely through stable agriculture and food supply thereby boosting production as water availability for crops will no longer be an issue. For this, the restoration or rehabilitation process has to be modulated toward consistent food production systems and pliant agricultural methods.
- Good health and well-being will be obtainable through refined nutrition, regulated by the type of crop. Air quality refinement would be achieved through shortened wind weathering from restored lands which can improve human health. As income generation would be guaranteed due to successful restoration of lands, mental health would benefit; moreover, a greener and healthier environment can be achieved.
- Quality education can be imparted to the human population through technical skill training and management practices which can be practically implemented

thereafter. In planning, management, and monitoring restoration projects, new technicalities are cropping up as mentioned by Kimiti et al. (2017).

Clean water and sanitation also benefit directly. Water percolating in the soil recharges underground water, and inhibited runoffs prevents flooding. As targeted, restoration and conservation of water ecosystems including lakes, aquifers, and rivers will hence be valuable.

6.4.2 A Blueprint for Land Revival and Spin-offs for Multiple SDGs

The following strategies outlined should be applicable disparately at each level keeping in view the variegated backgrounds including economic, environmental, cultural, and social references: (i) elaborate comprehensive and planned study, to identify potential collaborations and agreements, (ii) application of a panoramic approach to planning and execution – especially for landscapes with variable land potential, (iii) development of desirable explanation, and (iv) funding in areas that are likely to be sustainable.

6.4.2.1 The Elaborate Comprehensive and Planned Study, to Identify Potential Collaborations and Agreements

Before any resource commitment occurs, a systematic elaborate approach to pinpoint potential synergies is required. This helps in planning a cost-effective and optimal project which can identify new investments with higher returns than the one proposed earlier. The advanced systematic analysis also facilitates minimal fluky consequences and such exchanges to be avoided. Thus, any restoration project which impacts the merit, possession, or authority over land can directly or indirectly alter and influence SDGs to quite an extent. However, the analysis should involve these steps where both different collaborators and specialists have effective participation. A different outlook is important during the initial innovative phase to guarantee the inclusion of all inherent positive impacts and accord in the survey.

6.4.2.2 Application of a Panoramic Approach to Planning and Execution – Especially for Landscapes with Variable Land Potential

This guides how to evaluate land potential. Temporal disconnections occur due to prolonged restoration processes; however, minimization of the aforesaid can be done for SDGs, such as impoverishment and literacy by optimizing co-benefits. For instance, knowledge benefits can be surged by coaching and engaging inhabitants for work completion. Ecosystem service markets (IPBES 2018) address spatial disconnections where solitary, and a category of people in other areas bears the cost for restoration via payment for the desired service, including good caliber water from farmland-controlled watersheds to urban areas (Postel and Thompson 2005). For sustainably managing lands, certification programs also help (Blackman and Rivera 2011). Soil variability in fields and landscapes arrests soil disintegration and impacts the restoration of water percolation and flow wastage. Water percolation is altered due to degraded land, and infiltration is reduced. Although degradation is reversible to an extent by human intervention and degradation effect usually reduces, understanding soil variability can pinpoint vulnerable lands exposed to degradation. Soil variability helps understanding infiltration improvement scopes through soil organic matter and vegetation restoration. Different types of vegetation endure soils that are dissimilar, possibly increase infiltration and enhance the soil organic matter (UNEP 2016).

6.4.2.3 Develop Desirable Explanations

The topography helps to understand the diversity which promotes arbitration yield and the great return of investment. In this regard, success has been acquired in the areas of biological research rehabilitation via breeding programs, yielding plants species capable of stabilizing devastated lands and at the same time imparting nitrogen. New avenues open new areas for specific solutions involving soil microbiota. In this context, specific parts of degraded lands can be treated with microbes that are capable of pollution remediation, which would otherwise remain polluted for years (Tripathi et al. 2017). Soil stabilization and triggering soil biota are crucial in this regard.

6.4.2.4 Funding in Areas That Are Likely to Be Sustainable

In target restoration finding areas that would have persistent benefits remain a challenge. For promoting agroecological systems, climate change issues have to be addressed; however, the effects of land use on socioeconomic pressures in near future have little consideration. Specific land areas where interventions are critical and in places that are complementary to be abiding can be assessed to get the greatest short-term responses (Herrick et al. 2019). Review in this regard is mandatory for rehabilitation and productivity of land, or where land values are transposed due to new road formations. Figure 6.2 depicts proposed directions through which impoverished lives and land restoration may be upgraded.

Devastated land is a substantial hurdle for sustainable development leading to a yearly economic loss of approximately more than US $6 trillion. Thus, adopting consistent methods for the restoration of damaged land at global, local, and national levels is critical for prompt awareness toward UN-SDGs. As benefits and opportunities for land repair are granted by UN-DER, benefitting humans and surroundings, communities, people, and organizations can adopt these prospects to revive the conditions of disturbed lands for a sustainable future.

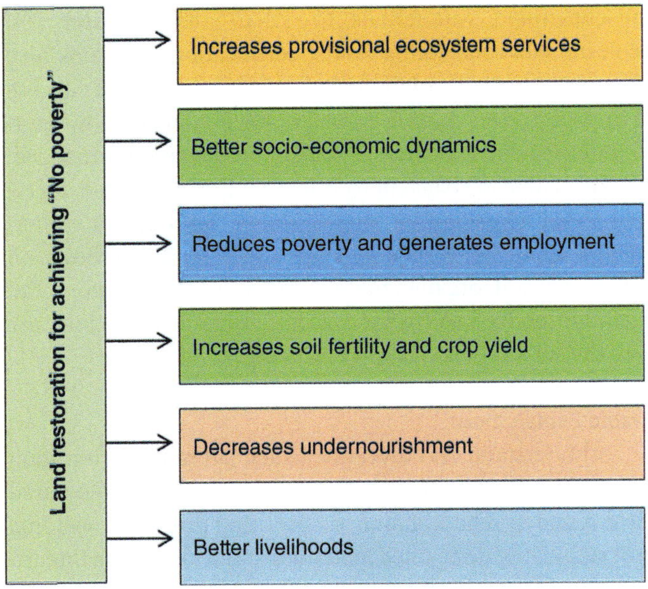

Figure 6.2 A notional depiction showing possible ways to poverty reduction and livelihood upliftment of the poor through land restoration. *Source:* Adapted from Cao et al. (2017).

6.5 Important Case Studies of Mine Restoration

The environment management execution in a resource sector should be prioritized at the end of any mining project to ensure proper rehabilitation. Planning and sequential strategy should be implemented which would take place step by step during the entire operation. Generally, some conventional approaches to mine land reclamation are well known. For example, use of laterite cover on waste discards, garland drain, hardy plants species like *Eucalyptus*, *Acacia* with the sole aim to develop a green capping and create biomass on the area without consideration to biodiversity or community use and post-mine closure land use is generally ignored. Unlike age-old traditions of mining, new emerging technologies, regulatory frameworks, and evolving corporate values have led the resource sectors to refine and develop innovative rehabilitation methods. This phenomenon can be termed progressive rehabilitation. These progressive approaches are afforestation – biotechnological approach, waste dyke reclamation – geo-textile approach, impoverished mine pits – backfilling approach, Agri-horticultural approach, pisciculture techniques, rain water harvesting, and aquatic sports. Considering progressive reclamation, some case studies on effective mine site restoration are discussed in the following:

Newlands, Glencore – is located in the northwest of the township of Glenden. It lies in the northern area of the Bowen Basin. Newlands has an open-cut, longwall mining operation, which produces coal and coking coal for exportation. During 2016 and 2017, in association with the Department of Environment and Heritage Protection (now known as the Department of Environment and Science), Glencore worked toward rehabilitation of the land and its surrounding zones, which are overburdened, toward creating ambient greenery with associated flora and fauna. Queensland Government certified and acknowledged Newlands as the first opencut mine to rehabilitate 73.48 hectares of land successfully. This was a significant move and provided immense scope for encouraging both the government and the mining sector for further rehabilitation processes. Since then, the team achieved progressive certification for another 138 hectares of rehabilitation, later the Queensland Government promoted an earlier mined land (in the 1990s) for rehabilitation to support agriculture and pasture grazing.

Rolleston, Glencore – Located in the southern area of the Bowen Basin, Rolleston is located to the southeast of Emerald situated in Queensland. Being an opencast maneuver, it generates moderate ash thermal coal for exports. This area is active toward progressive rehabilitation and revamping of degraded mine land for pastures to support livestock grazing. In this context, rehabilitation is administered with indigenous tree species and herbage grass species. The Government of Queensland certified a major portion of about 220.6 hectares of rehabilitated land in June 2018, and further 166 hectares in November 2019, adding to Glencore's growing assets.

Norwich Park, BHP Mitsubishi Alliance – Norwich Park is an opencast mine providing hard, coking in Queensland. This site is located 25 km southeast of Dysart and 256 km southwest of Mackay in the Bowen Basin. This mining area has been abandoned after 2012 and has been placed for maintenance. Restoration and rehabilitation both during and after mining have been undertaken by BHP Mitsubishi Alliance. A series of steps including re-modeling the topography, ripping, ameliorating, topsoil smearing, and seeding to prompt restoration are been adopted to restore the pre-mining condition of the land. In due course of time, success in this area has been achieved and some parts of the land are now used as grazing grounds by adjacent stakeholders. The Government of Queensland certified 294 hectares of rehabilitated land in February 2019.

Oaky Creek, Glencore – Situated in the Bowen Basin of Central Queensland, Oaky Creek is 90 km northwest of Emerald and 46 km southwest of Middlemount. After 2006, opencast mining activities stopped; however, lower-level operations continued to cater to coal for metallurgical export. As earlier adopted measures of land revamping through sustainable vegetation species had been successful, similar measures were adopted here to promote cattle grazing. Finally, the Government of Queensland certified 132.8 hectares of rehabilitated land in December 2020. Taking a step up in this area, another 369 hectares of rehabilitation has started in 2020 for future certification applications.[9]

Orasso Dongor Mine – Orosso Dongor mine exists in the northern fringes of Goa and is extended to about 100 Ha. Mining activities were carried out during the time frame of the 1960s and exhausted in 1974. Fast-growing trees were the quick developers in the overburdened dumps, later cashew plantations were adopted which provides fiscal returns to local dwellers. At present, the entire dyke area has turned out into a scenic cashew orchard. Moreover, a natural fountain existing at the mine land is conserved by channelizing the water and saving it as a potable water source for adjacent villages.

- *Sanquelim Mine, Goa, India* – Sanquelim commonly known as Sankhali is a municipal council town in North Goa district in the Indian state of Goa. To reclaim the devastated mine lands, a scientific approach for reclamation is undertaken in collaboration with organizations like "FRI, NIO, State Agriculture Department, State Forest Department, Goa University." The exhausted mine conversion into an Eco-tourist Spot has already been given nod by the management and concerned regulatory bodies. Moreover, an arboriculture approach has been adopted to develop Aromatic Garden, Spice plantation, Medicinal garden, Nakshatra Vruksh Devta Udyan, Charak Vatika, Pisciculture pond, Butterfly garden, in this site. Considering the reclamation of the Mine Pit, in 1990, it was developed into a pisciculture pit in alliance with "NIO [National Institute of Oceanography] situated at Dona Paula, Goa." Water treatment was carried out before fish specimens were added, to check the nutrient status and pH in the pit measuring 150 m × 30 m having a depth of 6 m. Fish fingerlings of Rohu, Mrigal, and Carp were later released with critical monitoring of temperature and feed which showed positive growth results for pisciculture. This case study shows biotechnologically reclaimed mine sites before and after 10 years.[10]

6.5.1 Case Studies of Mine Rehabilitation from Australia

Australian mining companies believe in the essentials of land rehabilitation for responsible mining. Of late, mine rehabilitation is modulated, better administered, and more liable than ever before. This industry has a commendable approach toward land rehabilitation that has improved significantly over past decades.

- *Anglo American* – Initially used as a spoil dump and topsoil stockpile a 165-hectare pit was successfully rehabilitated in the Dawson coal mine in Central Queensland by the Anglo American. The revamping of pit started in 2012 to remodel land for local farming and livestock, which has benefitted cattle immensely and catered to their food needs. Anglo American committed to long-term rehabilitation it comes up with sustainable post-mining use. More than 400 hectares of mined land in 2020 was rehabilitated using advanced data technology similar to 740 football fields.

- *Yancoal Australia* – An opencast coal mine near Collie Western Australia, Yancoal's WO5B was operational from 1970s to 1996. Later, during the 1980s rehabilitation work began with larger ideas that changed over years due to alternate weather, mine extensions, and change in legislation. However, Yancoal's commitment toward sustainable rehabilitation of the area molded the land into a state that benefited both the environment and the community. Thus, by December 2020, Yancoal was successful in land rehabilitation and Lake Kepwari was created in association with the state government and the local community to provide a 100-hectare aquatic tourist attraction.
- *Newmont* – After Newmont possessed the non-functional "Woodcutters lead–zinc mine in the Northern Territory" in the 2002 acquisition of Normandy, Newmont has participated in the rehabilitation of inactive mines in collaboration with the aboriginals of that area, the Kungarakan, and Warai people. The activities are monitored as per the agreement status of Woodcutters, which specifies employment generation for natives, pedagogy, and assurance from allies to return the land to the original holder after termination benchmarks are fulfilled. The present case study shows the status of Newmont Woodcutters lead–zinc mine post-rehabilitation.
- *Peabody* – After the completion of coal mining in 2013, Wilkie Creek has accomplished recuperation of its Wilkie Creek site in Queensland with more than 60% of rehabilitation completed. Refilling of opencast pits, remodeling of dykes, and leveling-related works were mainly adopted in the project. The ultimate landscape design includes pastures and livestock troughs to bear the final use of grazing. Substantial group engagement in post-mine rehabilitation through backfilling along with grazing trials has furnished constructive results for adjacent graziers. This mine case study depicts Wilkie Creek rehabilitation over the years 2008–2016.

6.6 Policy Restructuring for Fusing Circular Economy and Mine Land Restoration

Every year, mining assignments impart large volumes of mining waste. On an average, currently 3500 functional mining refuse provisions globally (as mentioned by Mining, Minerals, and Sustainable Development Project [MMSD]), having barren rock dumps and tailing dams (Rankin 2015). Huge piles of mining wastes need improvised management both temporary as well as enduring (Adiansyah et al. 2015). Moreover, to fulfill the increasing metal demands, extraction of low-grade ores is in practice which adds to the tailings (Chryss et al. 2012). Improper management of such wastes can lead to massive pollution of air and water (through leaching of AMD); moreover, it can cause a calamity if not controlled (Lèbre and Corder 2015). For effective mine waste management, powerful technological means can be adopted for underground mining rather than opting for open-pit mining (Wellmer et al. 2019). Rethinking mine waste management is crucial which requires effective inputs from stakeholders to remodel current operations and future projects which would deliver cleaner products and re-engineer by-products in closed-loop systems (Rankin 2015). In this regard, the perception of a circular economy was conceived to develop a sphere with no wastes. As formulated by Kirchherr et al. (2017) circular economy is a financial system developed on the basis of a trading strategy which does not encourage wastage of any product but emphasizes on the concept of reducing, recycling, or remodelling of commodities and products and their utilization at consumer, eco-industrial, or national/global level to initiate sustainable development to ensure future financial, environmental, and social prosperity for the coming generations.

Earlier, the linear model was a structure where input materials were pocket friendly, handy, and reachable but are no longer continued and persistent. Thus, the need for a circular model is developed which is derived over time after referring to multiple findings and not confined to any single author.

For resource-deficient overpopulated cities, circular practices are viable as they give optimal benefits in terms of cost savings. Moreover, policy changes are easier at city levels, compared to state or national ones where administrative timelines and guidelines may impede the methodical implementation of innovative circular concepts. Thus, at city levels, pilot initiatives are easier to execute as adaptation is quick and less challenging. As Ellen MacArthur Foundation has defined, "a circular city embeds the principles of a circular economy across all of its functions, establishing an urban system that is regenerative and restorative by design." In these cities, no waste is discarded and is utilized at its maximum using technologies and improvised logistics. Such a city aims to bring about affluence and budget flexibility for itself and its inhabitants. A circular economy is driven by three basics: elimination, circulation, and regeneration. Figure 6.3 represents the basics of circular economy, and Figure 6.4 depicts circular economy transition.

Certain circular economy principles and guidelines (Figure 6.4) can help in creating a definite perspective or outline a road map to imply circularity in urban areas. They are:

- Closed-loop – all matter enters into a boundless cycle (technical or biological).
- Emission reduction – renewable sources are the ultimate energy source.
- Value generation – assets are used to create value.
- Modular design – modular and alterable design of products helps better resilience of framework.
- Creative business prototype – new ideas enable the shifts from possession to use of services.

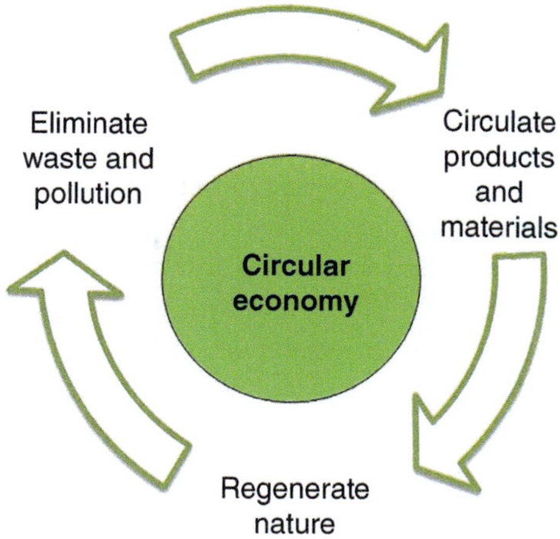

Figure 6.3 The basics of a circular economy.

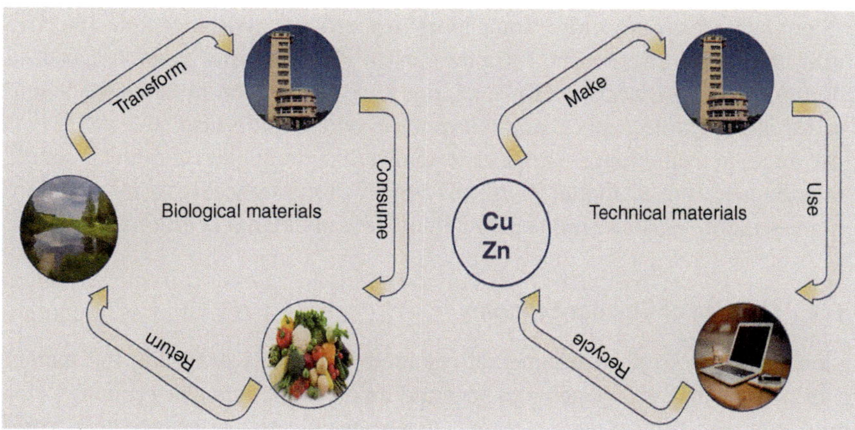

Figure 6.4 Circular economy transition.

- Region-based reverse logistics – an area identified service with product recovery capabilities.
- Natural systems upgradation – human intervention promotes ecological community and associated services, and the replotting of "natural assets." (Blok et al. 2016; Circle Economy, TNO, and Fabric 2016).

A circular city motivates administration thinking to facilitate lucrative, community, and benefits of the nature along with a polished life quality for its citizens. Innovative circular ideas and methods, induced with contemporary and bureaucratic formats, should be adopted by nations to consider circularity as mentioned in the Sustainable Development Goals for 2030. Some of the circular solutions can include:

- Urbanization – Rapid urban growth increases pressure on the environment and its resources, thus stretching limited resources and areas expands to harbor greater business and community.
- Stock and cost risks – Urban economies can be affected by rising prices of raw commodities and disruptions; however, adopting circular techniques can reduce the risk by enhancing the market for stock items.
- Ecosystem deterioration – Most liquid, solid, and organic wastes from urban development and landfills cause overburdening of ecosystems through degraded climate and loss of biodiversity. This can be reduced through circular practices and solutions.
- Environmental liability – Markets and administration are increasingly aware of intensified demands for liability and the hazards to brands from an unpleasant ecological impression.
- Client behavior – Innovative trade ideas (e.g. product assistance and collective utilization) can mitigate the extra burden on waste removal processes, caused due to over consumerism. However, this experimentation in conveyance and sociability is gaining appeal and relevance to other sectors also.
- Advances in technology – Large-scale application of circular economy practices is enhanced due to digital platforms which promote access to information, management, logistics, and facility deployment of circular solutions.

6.6.1 Mileage of Circular Economy

Rational utilization of environmental resources as well as reviewing the market trends are essential components to generate flux toward a circular economy. This would imply changing customer attitude, synchronizing production processes. Apart from inspiring new livelihood prospects and diminishing crude demands, this strategy would modify customer behavior, production, and regulation processes too.

Thus net gains through circular economy would create:

- A circular supply chain, where the by-products of one are easily managed into another process, introducing new values from the parent material.
- Increased credibility of work and improvised usage of shared products.
- Keeping in view the concept of sustainability, more capable products either repaired or reused from earlier materials benefit all stakeholders.
- People can gain financially through the expanded lifespan of products. Revenue can be generated from unused and remaining materials also reducing their ecological footprint, thus ensuring continuance, stocking, servicing, and elimination costs for customers.

- A circular economy creates openings for financial expansion toward new avenues such as collection and reverses plannings, revending, constituent remodeling, and facelift. Curtailing the use of pure substances lowers valuation swings with susceptibility to price inconsistency and stock disturbance due to catastrophic or bureaucratic variation. Circular economy implementation promotes a broad reconsideration of supply chains, which can lead to improved adaptability. Commodity as a service archetype involves extended relationships between consumer and distributor, giving ample scope to improve customer delight and commitment as well as improved marquee reputation by viable practices (Ellen MacArthur Foundation, 2015).
- Through a circular economy, excess resource depletion is lessened facilitating cities to fulfill profitability and minimizing negative environmental influences caused by waste incineration and/or dumping wastes. This can create new occupations in areas such as recycling and renovation practices and coordination.

6.7 Conclusion

The core of rehabilitation is judged based on the sustainability of natural capital and not solely on the cost benefits and technological proofs. This marks the sustainability of present and future generations too. Based on the available knowledge of fixing and mending current scenarios, materially and commercially, permission for upcoming projects should be granted. Policy makers and stakeholders have an important role in outlining rehabilitation standards, improvement prospects, and approaches. They are also entitled to design powerful metrics to measure performances against normals and document them. Restoring ecosystem services is not possible automatically, but mine land reclamation does add up to the improvement of ecosystem services. Ecosystem service in environmental assessments needs two essentials, amalgamation of social and environmental issues and the involvement of communities (Rosa and Sánchez 2015). Additionally, the land remodeling program should target the most important ecosystem services to communities.

Notes

1 https://www.minerals.org.au/mine-rehabilitation-case-studies
2 https://americanmineservices.com/land-rehabilitation-after-mining/
3 https://www.motherearthnews.com/sustainable-living/nature-and-environment/organic-lavender-farm-reclaims-land-lives-economy-zbcz2108
4 www.undp.org

5 https://www.resourcepanel.org/reports/land-restoration-achieving-sustainable-development-goals
6 www.foreststreesagroforestry.org
7 www.sdgs.un.org
8 www.un.org
9 https://www.qrc.org.au/rehabilitation-case-studies/
10 http://cdn.cseindia.org/userfiles/Mine-Reclamation-&-Restoration-at-Sesa-Goa-Mahesh-Patil.pdf)

References

Abhilash, P.C., Tripathi, V., Edrisi, S.A. et al. (2016). Sustainability of crop production from polluted lands. *Energy, Ecology and Environment 1* (1): 54–65.

Adiansyah, J.S., Rosano, M., Vink, S., and Keir, G. (2015). A framework for a sustainable approach to mine tailings management: disposal strategies. *Journal of Cleaner Production 108*: 1050–1062.

Ahirwal, J. and Maiti, S.K. (2016). Assessment of soil properties of different land uses generated due to surface coal mining activities in tropical Sal (Shorea robusta) forest, India. *Catena 140*: 155–163.

Ahirwal, J. and Maiti, S.K. (2018). Development of Technosol properties and recovery of carbon stock after 16 years of revegetation on coal mine degraded lands, India. *Catena 166*: 114–123.

Ahirwal, J., Maiti, S.K., and Singh, A.K. (2016). Ecological restoration of coal mine-degraded lands in dry tropical climate: what has been done and what needs to be done? *Environmental Quality Management 26*: 25–36. https://doi.org/10.1002/tqem.21481.

Ali, H., Khan, E., and Sajad, M.A. (2013). Phytoremediation of heavy metals—concepts and applications. *Chemosphere 91* (7): 869–881.

Aronson, J., Blignaut, J.N., Milton, S.J. et al. (2010). Are socioeconomic benefits of restoration adequately quantified? A meta-analysis of recent papers (2000–2008) in Restoration Ecology and 12 other scientific journals. *Restoration Ecology 18* (2): 143–154.

Asher, C.J., Grundon, N.J., and Menzies, N. (2002). How to unravel and solve soil fertility problems. ACIAR Monograph No. 83. Australian Centre for International Agricultural Research, Canberra, Australia, 139pp. ISBN 1 86329 321 4 (print).

Bandyopadhyay, S. and Maiti, S.K. (2019). Evaluation of ecological restoration success in mining-degraded lands. *Environmental Quality Management 29* (1): 89–100.

Baral, H., Guariguata, M.R., and Keenan, R.J. (2016). A proposed framework for assessing ecosystem goods and services from planted forests. *Ecosystem Services 22*: 260–268.

Barr, S., Jonas, J.L., and Paschke, M.W. (2017). Optimizing seed mixture diversity and seeding rates for grassland restoration. *Restoration Ecology 25* (3): 396–404.

Bell, L.C. (2002). Remediation of chemical limitations. In: *Restoration and management of derelict land—modern approaches* (ed. M.H. Wong and A.D. Bradshaw), 112–127. Singapore: World Scientific Publishing Co.

Benayas, J.M.R., Newton, A.C., Diaz, A., and Bullock, J.M. (2009). Enhancement of biodiversity and ecosystem services by ecological restoration: a meta-analysis. *Science 325* (5944): 1121–1124.

Blackman, A. and Rivera, J. (2011). Producer-level benefits of sustainability certification. *Conservation Biology* 25 (6): 1176–1185.

Blok, K., Hoogzaad, J., Ramkumar, S. et al. (2016). *Implementing Circular Economy Globally Makes Paris Targets Achievable*. Circle Economy, Ecofys.

Bradshaw, A. (2000). The use of natural processes in reclamation—advantages and difficulties. *Landscape and Urban Planning 51* (2–4): 89–100.

Bryce, R., Irvine, K.N., Church, A. et al. (2016). Subjective wellbeing indicators for large-scale assessment of cultural ecosystem services. *Ecosystem Services 21*: 268–269.

Bullock, J.M., Aronson, J., Newton, A.C. et al. (2011). Restoration of ecosystem services and biodiversity: conflicts and opportunities. *Trends in Ecology & Evolution 26* (10): 541–549.

Bullock, C., Joyce, D., and Collier, M. (2018). An exploration of the relationships between cultural ecosystem services, socio-cultural values and well-being. *Ecosystem Services 31*: 142–152.

Cao, S., Shang, D., Yue, H., and Ma, H. (2017). A win-win strategy for ecological restoration and biodiversity conservation in Southern China. *Environmental Research Letters 12* (4): 044004.

Catacutan, D.C. and Villamor, G.B. (2016). Gender roles and land use preferences—implications to landscape restoration in Southeast Asia. In: *Land Restoration* (ed. I. Chabay, M. Frick and J. Helgeson), 431–440. Academic Press ISBN 978-0-12-801231-4, https://doi.org/10.1016/C2013-0-18711-9.

Chan, K.M. and Satterfield, T. (2016). Managing cultural ecosystem services for sustainability. In: *Routledge Handbook of Ecosystem Services* (ed. M. Potschin, R. Haines-Young, R. Fish and R.K. Turner), 343–358. Routledge eBook ISBN9781315775302, https://doi.org/10.4324/9781315775302.

Chrysargyris, A., Michailidi, E., and Tzortzakis, N. (2018). Physiological and biochemical responses of Lavandula angustifolia to salinity under mineral foliar application. *Frontiers in Plant Science 9*: 489.

Chryss, A., Fourie, A.B., Monch, A. et al. (2012). Towards an integrated approach to tailings management. *Journal of the Southern African Institute of Mining and Metallurgy 112* (11): 965–969.

Clements, W.H., Vieira, N.K., and Church, S.E. (2010). Quantifying restoration success and recovery in a metal-polluted stream: a 17-year assessment of physicochemical and biological responses. *Journal of Applied Ecology 47* (4): 899–910.

Cortinovis, C. and Geneletti, D. (2018). Ecosystem services in urban plans: what is there, andwhat is still needed for better decisions. *Land Use Policy 70*: 298–312.

Cowie, A.L., Orr, B.J., Sanchez, V.M.C. et al. (2018). Land in balance: the scientific conceptual framework for Land Degradation Neutrality. *Environmental Science & Policy 79*: 25–35.

Cristescu, R.H., Frère, C., and Banks, P.B. (2012). A review of fauna in mine rehabilitation in Australia: current state and future directions. *Biological Conservation 149* (1): 60–72.

Darvill, R. and Lindo, Z. (2015). Quantifying and mapping ecosystem service use across stakeholder groups: implications for conservation with priorities for cultural values. *Ecosystem Services 13*: 153–161.

Das, R. and Maiti, S.K. (2016). Importance of carbon fractionation for the estimation of carbon sequestration in reclaimed coalmine soils—a case study from Jharia coalfields, Jharkhand, India. *Ecological Engineering 90*: 135–140.

De Gruijter, J.J. (2002). Sampling. In: *Methods of Soil Analysis. Part 4. Physical Methods*. (ed. J.H. Dane and G.C. Topp), 45–80. Madison, Wisconsin: Soil Science Society of America, Inc.

Dollhopf, D.J. (2000). Sampling strategies for drastically disturbed lands. *Reclamation of Drastically Disturbed Lands 41*: 21–40.

Ellen MacArthur Foundation (2015). Delivering the Circular Economy: A Toolkit for Policymakers.

Elton, C.S. (1958). *The Ecology of Invasions by Animals and Plants*. London: Methuen.

Farber, S., Costanza, R., Childers, D.L. et al. (2006). Linking ecology and economics for ecosystem management. *Bioscience 56* (2): 121–133.

Gastauer, M., Trein, L., Meira-Neto, J.A.A., and Schumacher, W. (2013). Evaluation of biotope's importance for biotic resource protection by the Bonner Approach. *Ecological Indicators 24*: 193–200.

Geneletti, D. (2013). Assessing the impact of alternative land-use zoning policies on future ecosystem services. *Environmental Impact Assessment Review 40*: 25–35.

Geneletti, D. (2016). *Handbook on Biodiversity and Ecosystem Services in Impact Assessment*. Cheltenham, UK; Northampton, MA, USA: Edward Elgar Publishing.

Geneletti, D., Scolozzi, R., and Adem Esmail, B. (2018). Assessing ecosystem services and biodiversity tradeoffs across agricultural landscapes in a mountain region. *International Journal of Biodiversity Science, Ecosystem Services & Management 14* (1): 188–208.

Ghosh, D., Masto, R.E., and Maiti, S.K. (2020). Ameliorative effect of Lantana biochar on coal mine spoil and growth of maize (*Zea mays*). *Soil Use and Management 36* (4): 726–739. https://doi.org/10.1111/sum.12626.

Giannini, T.C., Giulietti, A.M., Harley, R.M. et al. (2017). Selecting plant species for practical restoration of degraded lands using a multiple-trait approach. *Austral Ecology 42* (5): 510–521.

Gilbart, M. (2012). *Under Cover: Wildlife of Shrublands and Young Forest*, 87p. Cabot, VT: Wildlife Management Institute http://youngforest.org/sites/default/files/Under_ Cover-010412_FINAL.pdf (accessed 1 October 2016).

Groninger, J., Skousen, J., Angel, P. et al. (2017). Mine reclamation practices to enhance forest development through natural succession. In: *The Forestry Reclamation Approach: Guide to Successful Reforestation of Mined Lands. General Technical Report. NRS-169* (ed. M.B. Adams), 1–7. Newtown Square, PA: *US Department of Agriculture, Forest Service, Northern Research Station*.

Hall, S.L., Barton, C.D., and Baskin, C.C. (2010). Topsoil seed bank of an oak-hickory forest in eastern Kentucky as a restoration tool on surface mines. *Restoration Ecology* 18 (6): 834–842. https://doi.org/10.1111/j.1526-100x.2008.00509.x.

Harris, J.A. and van Diggelen, R. (2006). Ecological restoration as a project for global society. In: *Restoration Ecology* (ed. J. Van Andel and J. Aronson), 3–15. Malden, MA, USA: *Blackwell Science*.

Harris, T.M., Zaimes, G.G., Khanna, V., and Landis, A.E. (2015). Sunflower cultivation on coal mine refuse piles in Appalachia for diesel biofuel production from a life-cycle perspective. *Procedia Engineering 118*: 869–878.

Hermes, J., Van Berkel, D., Burkhard, B. et al. (2018). Assessment and valuation of recreational ecosystem services of landscapes. *Ecosystem Services 31*: 289–295. https://doi.org/10.1016/j.ecoser.2018.04.011.

Herrick, J.E., Neff, J., Quandt, A. et al. (2019). Prioritizing land for investments based on short-and long-term land potential and degradation risk: a strategic approach. *Environmental Science & Policy 96*: 52–58.

Holl, K.D. and Aide, T.M. (2011). When and where to actively restore ecosystems? *Forest Ecology and Management 261*: 1558–1563.

Howe, C., Suich, H., Vira, B., and Mace, G.M. (2014). Creating win-wins from trade-offs? Ecosystem services for human well-being: a meta-analysis of ecosystem service trade-offs and synergies in the real world. *Global Environmental Change 28*: 263–275.

ICMM – International Council on Mining & Metals (2019). *Integrated Mine Closure Good Practice Guide*, 2ee. ICMM. https://www.icmm.com/en-gb/guidance/environmental-stewardship/2019/integrated-mine-closure.

IPBES (2018). *Summary for Policymakers of the Assessment Report on Land Degradation and Restoration of the Intergovernmental Science-Policy Platform on Biodiversity and Ecosystem Services* (ed. R. Scholes, L. Montanarella, A. Brainich, et al.), 44p. Bonn, Germany: *IPBES Secretariat*.

IUSS Working Group WRB (2006). World reference base for soil resources 2006. In: *World Soil Resources Reports No. 103*, 2ee. Rome: FAO. ISBN 92-5-105511-4.

Jacob, C., Vaissiere, A.C., Bas, B., and Calvet, C. (2016). Investigating the inclusion of ecosystem services in biodiversity offsetting. *Ecosystem Services 21*: 92–102. https://doi.org/10.1016/j.ecoser.2016.07.010.

Jaligot, R., Kemajou, A., and Chenal, J. (2018). Cultural ecosystem services provision in response to urbanization in Cameroon. *Land Use Policy 79*: 641–649.

Kharytonov, M.M., Klimkina, I., Martynova, N. et al. (2019). The biochar impact on miscanthus and sunflower growth in marginal lands. *Agrology, 2020* 3 (1): 3–11. https://doi.org/10.32819/020001.

Kimiti, D.W., Hodge, A.M.C., Herrick, J.E. et al. (2017). Rehabilitation of community-owned, mixed-use rangelands: lessons from the Ewaso ecosystem in Kenya. *Plant Ecology 218* (1): 23–37.

Kirchherr, J., Reike, D., and Hekkert, M. (2017). Conceptualizing the circular economy: an analysis of 114 definitions. *Resources, Conservation and Recycling 127*: 221–232.

Koch, J.M. and Hobbs, R.J. (2007). Synthesis: is Alcoa successfully restoring a jarrah forest ecosystem after bauxite mining in Western Australia? *Restoration Ecology 15*: S137–S144.

Kumar, A., Ahirwal, J., Maiti, S.K., and Das, R. (2015a). An assessment of metal in fly ash and their translocation and bioaccumulation in perennial grasses growing at the reclaimed opencast mines. *International Journal of Environmental Research 9* (3): 1089–1096.

Kumar, S., Maiti, S.K., and Chaudhuri, S. (2015b). Soil development in 2–21 years old coalmine reclaimed spoil with trees: a case study from Sonepur-Bazari opencast project, Raniganj Coalfield, India. *Ecological Engineering 84*: 311–324.

Kumari, S. and Maiti, S.K. (2019). Reclamation of coalmine spoils with topsoil, grass, and legume: a case study from India. *Environmental Earth Sciences 78* (14): 1–14.

Kumari, S., Ahirwal, J., and Maiti, S.K. (2022). Reclamation of industrial waste dump using grass-legume mixture: an experimental approach to combat land degradation. *Ecological Engineering 174*: 106443.

Larondelle, N. and Haase, D. (2012). Valuing post-mining landscapes using an ecosystem services approach—an example from Germany. *Ecological Indicators 18*: 567–574.

Lawrence, D.P. (2003). *Environmental Impact Assessment: Practical Solutions to Recurrent Problems*. Hoboken, NJ: John Wiley & Sons.

Lazoz-Chavero, E., Zinda, J., Bennett-Curry, A. et al. (2016). Stakeholders and tropical reforestation: challenges, trade-offs, and strategies in dynamic environments. *Biotropica 48* (6): 900–914.

Lèbre, É. and Corder, G. (2015). Integrating industrial ecology thinking into the management of mining waste. *Resources 4* (4): 765–786.

Li, Y., Wu, Z., Dong, X. et al. (2019). Variance in bacterial communities, potential bacterial carbon sequestration and nitrogen fixation between light and dark conditions under elevated CO_2 in mine tailings. *Science of The Total Environment 652*: 234–242.

Litvaitis, J.A. (1993). Response of early successional vertebrates to historic changes in land use. *Conservation Biology 7* (4): 866–873.

Maestre, F.T., Quero, J.L., Gotelli, N.J. et al. (2012). Plant species richness and ecosystem multifunctionality in global drylands. *Science 335* (6065): 214–218.

Maiti, S.K. (2007). Bioreclamation of coalmine overburden dumps—with special empasis on micronutrients and heavy metals accumulation in tree species. *Environmental Monitoring and Assessment 125* (1): 111–122.

Maiti, S.K. (2012). *Ecorestoration of the Coalmine Degraded Lands*. Springer Science & Business Media.

Maiti, S.K. (2013a). Biodiversity erosion and conservation in ecorestored site. In: *Ecorestoration of the Coalmine Degraded Lands*, 187–199. India: Springer.

Maiti, S.K. and Maiti, D. (2015). Ecological restoration of waste dumps by topsoil blanketing, coir-matting and seeding with grass–legume mixture. *Ecological Engineering 77*: 74–84.

Maiti, S.K., Kumar, A., and Ahirwal, J. (2016a). Bioaccumulation of metals in timber and edible fruit trees growing on reclaimed coal mine overburden dumps. *International Journal of Mining, Reclamation and Environment 30*: 231–244. https://doi.org/10.1080/17480930.2015.1038864.

Maiti, S.K., Kumar, A., Ahirwal, J., and Das, R. (2016b). Comparative study on bioaccumulation and translocation of metals in bermuda grass (Cynodon dactylon) naturally growing on fly ash lagoon and topsoil. *Applied Ecology and Environmental Research 14* (1): 112. https://doi.org/10.15666/aeer/1401_001012.

Mayer, M. and Woltering, M. (2018). Assessing and valuing the recreational ecosystem services of Germany's national parks using travel cost models. *Ecosystem Services 31*: 371–386.

Mert, Y. (2019). Contribution to sustainable development: re-development of post-mining brownfields. *Journal of Cleaner Production 240*: 118212. https://doi.org/10.1016/j.jclepro.2019.118212.

Metslaid, S., Stanturf, J.A., Hordo, M. et al. (2015). Response of Scots pine to climatic factors and management on the reclaimed areas of oil shale mining. *Environmental Science andPollution Research* https://doi.org/10.1007/s11356-015-5647-4.

Mukhopadhyay, S. and Maiti, S.K. (2010). Phytoremediation of metal mine waste. *Applied Ecology and Environmental Research 8* (3): 207–222.

Mukhopadhyay, S. and Masto, R.E. (2016). Carbon storage in coal mine spoil by Dalbergia sissoo Roxb. *Geoderma 284*: 204–213.

Mukhopadhyay, S. and Masto, R.E. (2019). Effect of fly ash on carbon mineralization of biochar and organic manures added to mine spoil. *SN Applied Sciences 1*: 1119. https://doi.org/10.1007/s42452-019-1140-x.

Mukhopadhyay, S., Maiti, S.K., and Masto, R.E. (2013). Use of Reclaimed Mine Soil Index (RMSI) for screening of tree species for reclamation of coal mine degraded land. *Ecological Engineering 57*: 133–142.

Muñoz-Rojas, M., Román, J.R., Roncero-Ramos, B. et al. (2018). Cyanobacteria inoculation enhances carbon sequestration in soil substrates used in dryland restoration. *Science of the Total Environment 636*: 1149–1154.

Native Plant Network (2016). Native plants propagation protocol database. http://nativeplantnetwork.org. (accessed 30 November 2016).

Nussbaumer, Y., Cole, M.A., Offler, C.E., and Patrick, J.W. (2016). Identifying and ameliorating nutrient limitations to reconstructing a forest ecosystem on mined land. *Restoration Ecology 24* (2): 202–211.

Ohsowski, B.M., Dunfield, K., Klironomos, J.N., and Hart, M.M. (2018). Plant response to biochar, compost, and mycorrhizal fungal amendments in post-mine sandpits. *Restoration Ecology 26*: 63–72. https://doi.org/10.1111/rec.12528.

Orr, B.J., Cowie, A.L., Castillo Sanchez, V.M. et al. (2017). Scientific conceptual framework for land degradation neutrality. In: *A Report of the Science-Policy Interface*, 1–98. Bonn, Germany: *United Nations Convention to Combat Desertification (UNCCD).*

Orsi, F. and Geneletti, D. (2013). Using geotagged photographs and GIS analysis to estimate visitor flows in natural areas. *Journal for Nature Nonservation 21* (5): 359–368.

Palmer, M.A., Bernhardt, E.S., Schlesinger, W.H. et al. (2010). Mountaintop mining consequences. *Science 327* (5962): 148–149.

Pandey, V.C., Singh, K., Singh, J.S. et al. (2012). Jatropha curcas: a potential biofuel plant for sustainable environmental development. *Renewable and Sustainable Energy Reviews 16* (5): 2870–2883.

Peng, S.L., Liu, J., and Lu, H. (2005). Characteristics and role of Acacia auriculiformis on vegetation restoration in lower subtropics of China. *Journal of Tropical Forest Science 17* (4): 508–525.

Perring, M.P., Standish, R.J., Price, J.N. et al. (2015). Advances in restoration ecology: rising to the challenges of the coming decades. *Ecosphere 6* (8): 1–25.

Petrova, S. and Marinova, D. (2015). Using 'soft'and 'hard'social impact indicators to understand societal change caused by mining: a Western Australia case study. *Impact Assessment and Project Appraisal 33* (1): 16–27.

Pitz, C., Mahy, G., Vermeulen, C. et al. (2016). Developing biodiversity indicators on a stakeholders' opinions basis: the gypsum industry Key Performance Indicators framework. *Environmental Science and Pollution Research 23* (14): 13661–13671.

Polizzi, C., Simonetto, M., Barausse, A. et al. (2015). Is ecosystem restoration worth the effort? The rehabilitation of a Finnish river affects recreational ecosystem services. *Ecosystem Services 14*: 158–169.

Postel, S.L. and Thompson, B.H. Jr. (2005, May). Watershed protection: capturing the benefits of nature's water supply services. *Natural Resources Forum 29* (2): 98–108. https://doi.org/10.1111/j.1477-8947.2005.00119.x.

Prach, K. and Hobbs, R.J. (2008). Spontaneous succession versus technical reclamation in the restoration of disturbed sites. *Restoration Ecology 16*: 363–366.

Prach, K. and Pyšek, P. (2001). Using spontaneous succession for restoration of human-disturbed habitats: experience from Central Europe. *Ecological Engineering 17* (1): 55–62.

Rana, V. and Maiti, S.K. (2018). Differential distribution of metals in tree tissues growing on reclaimed coal mine overburden dumps, Jharia coal field (India). *Environmental Science and Pollution Research 25* (10): 9745–9758.

Ranasinghe, R.A.S.N., Maduwanthi, S.D.T., and Marapana, R.A.U.J. (2019). Nutritional and health benefits of jackfruit (Artocarpus heterophyllus Lam.): a review. *International Journal of Food Science* 2019: 4327183. https://doi.org/10.1155/2019/4327183.

Ranjan, V., Sen, P., Kumar, D., and Sarsawat, A. (2015). A review on dump slope stabilization by revegetation with reference to indigenous plant. *Ecological Processes 4* (1): 1–11.

Rankin, W.J. (2015). Towards zero waste. *AusIMM Bulletin 2015*: 32–37.

Raum, S. (2018). A framework for integrating systematic stakeholder analysis in ecosystem services research: stakeholder mapping for forest ecosystem services in the UK. *Ecosystem Services 29*: 170–184.

Richards, D.R. and Tunçer, B. (2018). Using image recognition to automate assessment of cultural ecosystem services from social media photographs. *Ecosystem Services 31*: 318–325.

Robiglio, V. and Reyes, M. (2016). Restoration through formalization? Assessing the potential of Peru's Agroforestry Concessions scheme to contribute to restoration in agricultural frontiers in the Amazon region. *World Development Perspectives 3*: 42–46.

Rosa, J.C.S. and Sánchez, L.E. (2015). Is the ecosystem service concept improving impact assessment? Evidence from recent international practice. *Environmental Impact Assessment Review 50*: 134–142.

Rosa, J.C.S. and Sánchez, L.E. (2016). Advances and challenges of incorporating ecosystem services into impact assessment. *Journal of Environmental Management 180*: 485–492.

Rosa, J.C.S., Sánchez, L.E., and Morrison-Saunders, A. (2018). Getting to "agreed" post-mining land use – an ecosystem services approach. *Impact Assessment and Project Appraisal 36* (3): 220–229. https://doi.org/10.1080/14615517.2018.1445175.

Rozas-Vásquez, D., Fürst, C., Geneletti, D., and Almendrae, O. (2018). Integration of ecosystem services in strategic environmental assessment across spatial planning scales. *Land Use Policy 71*: 303–310.

Sánchez, L.E., Silva-Sánchez, S.S., and Neri, A.C. (2014). *Guide for Mine Closure Planning*. Instituto Brasileiro de Mineração: *Brasilia, Brazil.*

SER [Society for Ecological Restoration] (2004). *The SER International Primer on Ecological Restoration. Version 2*. Tucson: Society for Ecological Restoration.

Sheoran, V., Sheoran, A.S., and Poonia, P. (2010). Soil reclamation of abandoned mine land by revegetation: a review. *International Journal of Soil, Sediment and Water 3* (2): 13.

Shrivastava, S. (2007). Tribal dependence on fly ash in Korba. *Journal of Ecological Anthropology 11* (1): 69–73.

Shukla, A.K. and Misra, P.N. (1993). Improvement of sodic soil under tree cover. *Indian Forester 119* (1): 43–52.

Skousen, J. (2010). *Revegetation Species and Practices*, 122–460. Virginia Cooperative Extension, Publication.

Stocking, M. (2001). *Land Degradation. International Encyclopedia of the Social & Behavioral Sciences*, 8242–8247. Amsterdam, Netherlands: Elsevier.

Suding, K.N. (2011). Toward an era of restoration in ecology: successes, failures, and opportunities ahead. *Annual Review of Ecology, Evolution, and Systematics 42*: 465–487.

Suding, K., Higgs, E., Palmer, M. et al. (2015). Committing to ecological restoration. *Science 348* (6235): 638–640.

Syktus, J.I. and McAlpine, C.A. (2016). More than carbon sequestration: biophysical climate benefits of restored savanna woodlands. *Scientific Reports 6* (1): 1–11.

Thompson-Hall, M. (2016). Land restoration, agriculture, and climate change: enriching gender programming through strengthening intersectional perspectives. In: *Land Restoration* (ed. I. Chabay, M. Frick and J. Helgeson), 421–430. Academic Press.

Tripathi, V., Edrisi, S.A., Chen, B. et al. (2017). Biotechnological advances for restoring degraded land for sustainable development. *Trends in Biotechnology 35* (9): 847–859.

UNEP (2015). *Policy Coherence of the Sustainable Development Goals. A Natural Resource Perspective*. Nairobi, Kenya: United Nations Environment Programme.

UNEP (2016). *Unlocking the Sustainable Potential of Land Resources: Evaluation Systems, Strategies and Tools. A Report of the Working Group on Land and Soils* (ed. J.E. Herrick, O. Arnalds, B. Bestelmeyer, et al.), 89pp. UNEP. http://www.resourcepanel.org/reports/unlocking-sustainable-potential-landresources (accessed 03/09/2022).

Ungaro, F., Häfner, K., Zasada, I., and Piorr, A. (2016). Mapping cultural ecosystem services: connecting visual landscape quality to cost estimations for enhanced services provision. *Land Use Policy 54*: 399–412.

Ussiri, D.A., Guzman, J.G., Lal, R., and Somireddy, U. (2019). Bioenergy crop production on reclaimed mine land in the North Appalachian region, USA. *Biomass and Bioenergy 125*: 188–195.

Walker, L.R. (1999). *Ecosystems of Disturbed Ground*. Elsevier.

Walker, L.R., Hölzel, N., Marrs, R. et al. (2014). Optimization of intervention levels in ecological restoration. *Applied Vegetation Science 17*: 187–192.

Wellmer, F.W., Buchholz, P., Gutzmer, J. et al. (2019). *Raw Materials for Future Energy Supply. Springer International Publishing.*

Whisenant, S. (1999). *Repairing Damaged Wildlands: A Process-orientated, Landscape-scale Approach*, vol. 1. Cambridge University Press.

Wilson, M.J. (2003). Book review on Dixon, J. B. & Schulze, D. G. (Editors) Soil Mineralogy with Environmental Applications. Published by the Soil Science Society of America. 866pp. Price $99.00. *Clay Minerals 38* (2): 255–257.

Yang, Y., Liang, Y., Ghosh, A. et al. (2015). Assessment of arbuscular mycorrhizal fungi status and heavy metal accumulation characteristics of tree species in a lead–zinc mine area: potential applications for phytoremediation. *Environmental Science and Pollution Research 22* (17): 13179–13193.

Yates, S.R. and Warrick, A.W. (2002). Geostatistics. In: *Methods of Soil Analysis. Part 4. Physical Methods* (ed. J.H. Dane and G.C. Topp), 81–118. Madison, Wisconsin: *Soil Science Society of America, Inc.*

Yurui, L., Xuanchang, Z., Zhi, C. et al. (2021). Towards the progress of ecological restoration and economic development in China's Loess Plateau and strategy for more sustainable development. *Science of the Total Environment 756*: 143676. https://doi.org/10.1016/j.scitotenv.2020.143676.

Zipper, C.E., Burger, J.A., Skousen, J.G. et al. (2011). Restoring forests and associated ecosystem services on appalachian coal surface mines. *Environmental Management 47* (5): 751–765.

Index

Eco-Restoration of Mine Land, First Edition. Vimal Chandra Pandey, Jitendra Ahirwal, Roopali Roychowdhury, and Ritu Chaturvedi.

d

Printed and bound by CPI Group (UK) Ltd, Croydon, CR0 4YY

17/09/2026

14964253-0001